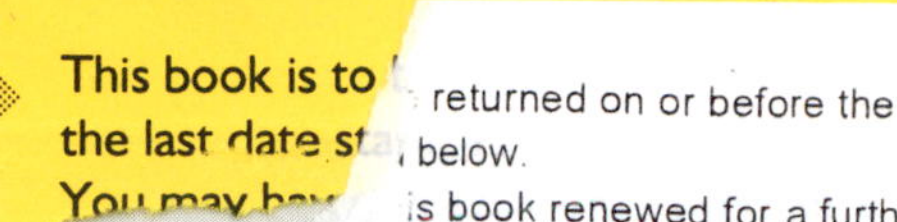

EXPERIMENTAL DESIGN AND ANALYSIS IN ANIMAL SCIENCES

EXPERIMENTAL DESIGN AND ANALYSIS IN ANIMAL SCIENCES

T.R. MORRIS

University of Reading
Department of Agriculture
Reading
UK

CABI *Publishing*

CABI *Publishing* is a division of CAB *International*

CABI Publishing
CAB International
Wallingford
Oxon OX10 8DE
UK

Tel: +44 (0)1491 832111
Fax: +44 (0)1491 833508
Email: cabi@cabi.org

CABI Publishing
10 E 40th Street
Suite 3203
New York, NY 10016
USA

Tel: +1 212 481 7018
Fax: +1 212 686 7993
Email: cabi-nao@cabi.org

A catalogue record for this book is available from the British Library, London, UK.
A catalogue record for this book is available from the Library of Congress, Washington DC, USA.

ISBN 0 85199 349 4

Typeset in Times by Columns Design Ltd, Reading, UK.
Printed and bound in the UK by Biddles Ltd, Guildford and King's Lynn.

Contents

Preface	x
A Note to Students	xi
Symbols and Acronyms	xii
1 Introduction	**1**
The Principles of Good Experiments	1
Randomization	2
Local Control	3
Summary	3
2 Blocking	**4**
Allocating Animals to Blocks and Treatments	7
Improvement in Precision Due to Blocking	9
Double Blocking	10
Confounding	13
Block × Treatment Interaction	15
Summary	16
Exercise 2.1	16
Exercise 2.2	17
3 Separating Treatment Means	**19**
Statistical Problems with Student's '*t*'-test	20
The Philosophy of Separating Treatment Means	20
An Appropriate Use of Multiple Range Testing	23
The Meaning of 'Significance'	23
Proving Two Things Equal	25
The Balance of Probability Argument	26
Dose–Response Trials	27
Summary	29
References	29
Exercise 3.1	29
Exercise 3.2	30

4 How Many Animals? 31
Estimating the CV of Future Experimental Material 32
Estimating the Difference to be Expected 34
Applying the Equation 36
Some Examples 36
What Are the Chances of Success? 37
Tabulation of Number of Replicates 38
What to Do if There Are Not Enough Animals 39
Summary 40
Exercise 4.1 40
Exercise 4.2 41
Exercise 4.3 41

5 Change-over Designs 42
Latin Squares 43
Balanced Latin Squares 46
Balanced Latin Squares with an Extra Period 48
Switchback Designs 49
When to Use Change-over Designs 50
Summary 51
Reference 51
Exercise 5.1 52

6 Pens and Paddocks 53
Groups of Animals in Pens 53
Keeping Records of Individuals 55
Grazing Trials 57
Coefficients of Variation for Groups 59
Summary 60
Exercise 6.1 60

7 Factorial Designs 62
Factorial Analyses with No Interactions 63
Factorial Analyses When Interaction Is Present 64
Two for the Price of One 64
Factorial Designs with Unequal Replication (Split Plots) 66
Summary 68
Exercise 7.1 69

8 Assumptions Underlying the Analysis of Variance 70
Homogeneity of Variances 70
The Logarithmic Transformation 71
Testing for Homogeneity of Variance 72
Further Examples 73
Other Transformations 73

Additivity 74
Summary 75
Reference 76
Exercise 8.1 76

9 Dose–Response Trials 78
Shapes of Response Curves 79
Asymptotic Responses 81
Fitting Straight Lines as a Compromise 83
Simple Curves 86
Exponential and Inverse Polynomial Models 87
The Reading Model 88
Choice of Treatments 90
Response Surfaces 92
Summary 92
References 92
Exercise 9.1 93

10 Uses of Covariance Analysis 95
Covariance Adjustment Using Preliminary Variables 96
Multiple Covariance 100
Blocking versus Covariance 101
Covariance Adjustment as an Aid to Interpretation 102
Summary 105
Exercise 10.1 105

11 Unbalanced Designs 107
Missing Plots 108
Unbalanced Designs 108
Some Examples 109
Summary 111
Reference 111
Exercise 11.1 111
Exercise 11.2 111

12 Repeated Measures 113
Time Trends 115
Weighing Ruminant Animals 116
Summary 117
Reference 117
Exercise 12.1 117

13 Discrete Data 119
Snags with the χ^2 Test 120
Estimating the Expected Outcome 121

Summary 122
Reference 122

14 Multiple Experiments **123**
Planning Multi-location Experiments 123
Reviewing Multiple Experiments 127
Summary 133
References 133
Epilogue 134

Appendix 1 **135**
Random Numbers and How to Use Them 135
Appendix 2 **137**
Some Useful General Formulae 137
Useful Quick Approximations 137
Differences between Differences 138
Appendix 3 **139**
Answers to Exercise 2.1 Using a Pocket Calculator 139
Appendix 4 **141**
Answers to Exercise 2.1 Using SAS 141
Appendix 5 **144**
Answers to Exercise 2.2 Using a Pocket Calculator 144
Appendix 6 **147**
Answers to Exercise 2.2 Using SAS 147
Appendix 7 **148**
Separation of Treatment Means 148
Orthogonal Polynomials 149
Orthogonal Polynomials for Regression Analysis 152
Appendix 8 **154**
Answers to Exercise 3.1 154
Answers to Exercise 3.2 155
Appendix 9 **158**
Answer to Exercise 4.1 158
Answer to Part 1 of Exercise 4.2 158
Answer to Part 2 of Exercise 4.2 159
Answer to Exercise 4.3 160
Appendix 10 **161**
Answers to Exercise 5.1 Using a Pocket Calculator 161
Appendix 11 **165**
Answers to Exercise 5.1 Using SAS 165
Appendix 12 **167**
Example ANOVA of a Balanced Latin Square Design 167
Appendix 13 **171**
Answers to Exercise 6.1 171
Appendix 14 **175**
Answers to Exercise 7.1 175

Appendix 15 **180**
Bartlett's Test 180
Appendix 16 **183**
Answers to Exercise 8.1 183
Appendix 17 **185**
Answers to Exercise 9.1 185
Appendix 18 **188**
Example of Analysis of Covariance 188
Appendix 19 **191**
Answers to Exercise 10.1 191
Appendix 20 **194**
Answers to Exercise 11.1 194
Appendix 21 **196**
Answers to Exercise 11.2 196
Appendix 22 **197**
Answers to Exercise 12.1 197
Appendix 23 **199**
Example of Matrix Used to Fit Constants for Trials 199
Appendix 24 **201**
Table of χ^2 201
Appendix 25 **202**
F-ratio Tables 202
Appendix 26 **204**
Student's *t* 204

Index 205

Preface

This book arises from experience in teaching postgraduate students who are about to undertake experiments with farm animals. Such students often find that questions about the design and analysis of their experiments have not been fully answered by their earlier training in basic statistics. They will have been introduced to a set of tools, but have not yet gained the experience to judge which tool is appropriate to each set of circumstances. Experience is, of course, a valuable teacher, but one of the purposes of education is to save the pupil from wasting too much time and effort in repeating mistakes that could be avoided by acquiring a little knowledge in advance.

There are no novel principles in the text that follows: indeed, all the statistical techniques referred to can be found in any good undergraduate text on statistical methods in agriculture or experimental biology. The novelty of the presentation is that it uses examples drawn from animal experiments to illustrate general principles and thus makes the learning process easier for those beginning research into animal production and animal science.

The book makes no attempt to discuss the ethics of using animals for experiments. This is not because such issues are unimportant, but simply reflects a decision to exclude ethics from the scope of this particular book. It is the author's view that the use of animals in experiments should be justified in every case (as is required by law in some countries, including the United Kingdom). A part of that justification is that the number of animals used should be the minimum sufficient to test properly the hypothesis being investigated. In this connection the book seeks to instruct the student in avoiding the twin pitfalls of using too few animals to produce any clear conclusions or, on the other hand, using more replicates than are strictly necessary. Since both these errors (and many others discussed in detail in the following chapters) are common in papers published by animal scientists, it seems that better training in experimental design is a desirable objective in all graduate schools where instruction in animal research methodology is part of the curriculum.

T.R. Morris
Reading 1998

A Note to Students

If you intend to make use of this book, it is as well that two things should be made clear at the outset.

First, the book presumes that you have had training in statistics sufficient to make you familiar with terms such as 'standard deviation' and 'analysis of variance'. You may be a little vague about the precise methods of calculating these things, perhaps because you took your statistics course some time ago, but it is essential to an understanding of this book that you have some background training in elementary statistics and that you are able to use a set of notes or a textbook which will refresh your memory when need be. *Statistical Methods in Agriculture and Biology*, 2nd edn, by R. Mead, R.N. Curnow and A.M. Hasted, published by Chapman & Hall, is a suitable text.

Secondly, the book is designed to be used with a £10 pocket calculator which will automatically give you the sum of the squares of a series of numbers read in. You are recommended to work through the exercises given as a means of learning the structure of the analysis, even though, in your own research, you will probably make use of a much more powerful computer with a statistical package that does all the computations for you. To know which bits of a statistical package you should be using, it is essential to have an understanding of the analysis you intend to undertake and, for people who lack the abstract conceptual abilities of a mathematician, this understanding comes most readily by working through simple numerical examples.

Symbols and Acronyms

ANOVA	analysis of variance
CF	correction factor
CP	crude protein (N $\times$ 6.25)
CV	coefficient of variation
d.f.	degrees of freedom
DM	dry matter (in feed)
DOM	digestible organic matter
F	Fisher's ratio (comparing two variances)
GnRH	gonadotrophin releasing hormone
LH	luteinizing hormone
LS	Latin square (see Chapter 5)
LSD	least significant difference (see Appendix 2)
ME	metabolizable energy
m.s.	mean square (of deviations from a mean) = variance
OMD	organic matter digestibility
P	probability (of something being due to chance)
RCB	randomized complete block design
SAA	sulphur-containing amino acid
SED	standard error of the difference between two means (see Appendix 2)
SEM	standard error of a mean (see Appendix 2)
s.p.	sum of products (i.e. $\sum(x - \bar{x})(y - \bar{y})$)
s.s.	sum of squares (of deviations about a mean)
t	Student's t value as given in Appendix 26
VFA	volatile fatty acid
x	a variable
y	a second variable, often dependent on x
$\sum$	the sum of
*	significantly greater than the 5% probability value
**	significantly greater than the 1% probability value
***	significantly greater than the 0.1% probability value

Chapter 1

Introduction

This book is intended to help those who are contemplating the use of live animals in their research to design better experiments and to analyse them effectively so as to extract valid conclusions. Alexander Pope wrote that 'a little learning is a dang'rous thing' and that is undoubtedly true, especially of statistics. But after reading and understanding this book your learning should have advanced from the stage typically reached at the end of an undergraduate's introduction to statistics (*dangerously* little learning) to a more subtle appreciation of what can and cannot be done and what *should not* be done by an animal scientist with access to a range of standard computing packages which he or she is allowed to use without the intervention of a professional statistician.

If you are lucky enough to have access to a proper statistician then, until you have substantial confidence based on experience, you should always take your proposed design to him or her *before* you begin an experiment. Statisticians can also be very helpful in sorting out problems after the data have been collected, but much the most important thing is to get the design right in the first place. It has been said (by Dr Daphne Vince-Prue) that the purpose of statistical design is to make statistical analysis unnecessary! This aphorism is not quite true, but a well-designed experiment will often yield results leading to clear conclusions even before the data are analysed statistically.

The Principles of Good Experiments

There are only *three* principles of experimental design. They are *replication*, *randomization* and *local control*.

If I were to tell you that a farmer gave his sick cow a home-made medicine and that a week later she had got better, you would be rightly sceptical about this 'evidence'. That was not an experiment because there was only one treatment. Any experiment must have at least *two treatments* (one of which may, if we choose, be a control involving no treatment). Suppose that the farmer had two sick cows and he gave his medicine to one of them: she got better in a week and the other, untreated, cow took a month to recover. This is now an experiment but it is unreplicated and will not impress us greatly. However, if the farmer had six cows showing similar symptoms and he gave his medicine to three of them, who recovered in 5–10 days, while the other three cows, left untreated, took 30–50 days to get better, you may well think that this farmer's medicine deserves to be taken seriously. This is now a *replicated* experiment. The whole basis of science is that observations are repeatable, although when we are using biological material we have to allow for the effects of natural variation which clouds our observations. We know that cows, like people, sometimes recover from illness whether they are treated or not and so we demand controls (animals *not* treated) *and* replication before we are satisfied that the evidence indicates that the medicine is the cause of the early recovery of treated cows.

The question of how much replication is needed is a technical one which has a technical answer (see Chapter 4). One of the most important advantages of this book is that it will enable you to calculate how many replicates are necessary in the context of a particular experiment. You will not be surprised to learn that replication has to be increased when experimental material is highly variable but can be reduced if the anticipated responses to treatment are large.

The essence of a scientific experiment conducted with variable biological material is that we *judge differences between units treated differently in the light of observed variation in units treated alike.*

Randomization

If I were now to tell you that the farmer reported above looked at his six ailing cows and chose the three that appeared least affected to receive his medicine, you would immediately revise your opinion of his 'experiment'. You would rightly say that the question of which cows were to be treated should have been determined *at random*. This, like replication, is such a commonsense precaution against biased results that it does not need much elaboration as a principle. The question of how randomness is achieved in practice will be dealt with in Chapter 2.

In medical research, where patients are randomly allocated to different medicines or to a medicine and a *placebo* (that is a tablet, cream or solution looking exactly like the real medicine but lacking the active ingredient), it is common practice to conceal the information about which patient is receiving which treatment from both the patients themselves and from the doctors recording the results of the trial. This is called a 'double-blind' trial. This procedure is not commonly applied in animal experiments, on the grounds that the observations made are '*objective*' (e.g. weights of animals or their products). But when, for example, behavioural observations are made, there is a serious risk that *subjective* bias may creep in and you should then aim to guard against this by specifying that, wherever possible, the person making the observations does not know which animals are receiving which treatment. There are, however, cases where the treatments applied are apparent even to a casual observer (trials assessing grazing behaviour on different sown pastures, for example) and then it becomes important to make the observations as objective as possible by, for example, formally recording with a stop-watch the times that animals spend in defined activities.

Local Control

Local control means imposing some restriction on the random allocation of treatments to take account of known or supposed *initial differences* in the experimental material. Much the most common form of local control in formal experimentation is *blocking*, which is described fully in Chapter 2.

Another kind of local control, often used in survey work, is *stratification*. This is like blocking, except that the numbers are usually ragged. For example, you might be interested in mastitis in dairy cows and the correlation between various husbandry and hygiene practices and the incidence of the disease. For this purpose you decide to take a random sample of the farms in a study area (you cannot visit and study them all) but before drawing names and addresses from a hat containing names of all the farmers in the designated area, you may decide to classify farms according to the size of their dairy herds. You will then draw similar proportions (at random) from each size class and thereby ensure that your survey fairly samples the situation in small, medium and large herds, and thus gives a truer picture of mastitis in the region. In such cases the number of farms in each stratum of the survey will probably be different, and the number yielding reliable information will probably differ from the number intended at the outset. Blocking is rather like stratification but, because it is applied in a planned experiment, it is usually more regular in its features.

Blocking is the first subject tackled in this book because, although allocation to blocks logically comes after you have chosen your treatments and assessed the number of replicates required, it forms a convenient springboard from which to remind you of things that you almost certainly know about statistics, but may have temporarily forgotten.

Summary

1. 'A little learning is a dang'rous thing'. This means you should learn more about *statistics*.

2. The principles of experimental design are *replication*, *randomization* and *local control*. The first two of these are indispensable features of any scientific experiment. Local control (blocking, etc.) is an optional extra, but can greatly improve an experiment.

Chapter 2

Blocking

Animals are often arranged in 'blocks' before they are allocated to treatments for an experiment. The term 'blocking' comes from field crop experimentation (as do most of the technical terms in experimental design). A field experiment arranged in blocks is shown in Fig. 2.1. This is called a *randomized complete block* (RCB) design: 'randomized' because treatments have been allocated randomly to positions within each block; 'complete' because each block contains every treatment.[a]

The *purpose* of blocking, whether for crop experiments or in animal trials, is to *increase precision* by reducing the error variance. This we can expect to do if (but only if) there is substantial variation between experimental plots associated with the feature used for blocking. In the experiment shown in Fig. 2.1, it is likely that two plots in block I are more alike in their potential yield characteristics than two plots drawn from distant blocks (e.g. I and VI). To the extent that the six blocks do differ in their average yield more than plots within the same block, those differences will be isolated in the analysis of variance,

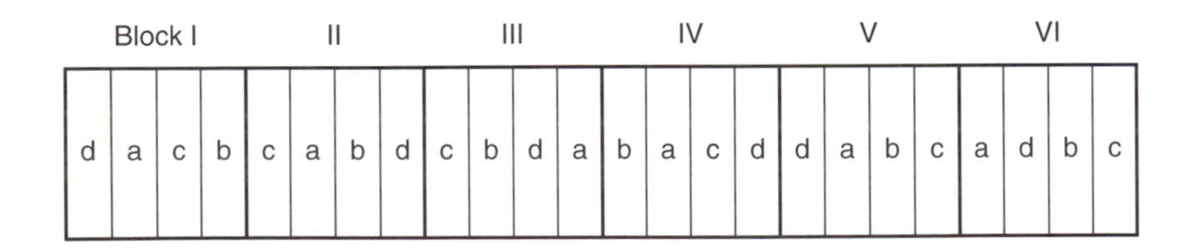

Fig. 2.1. A field experiment in which four treatments (a, b, c and d) are allocated at random within each of six blocks of land (I–VI).

[a] Incomplete block designs are seldom used in whole animal trials: their purpose is to allow blocks to contain fewer plots than the total number of treatments and so they are most useful when the number of treatments is large. It may be worth considering an incomplete block design if you are planning a change-over trial (see Chapter 5) in which there are more treatments than you can conveniently apply *successively* to a single animal.

thus reducing the error compared with what would have been observed if treatments had been allocated entirely at random.

There is, however, a price to pay for blocking. The *cost of blocking* is a loss of degrees of freedom (d.f.) from the error term. The structures of the analyses of variance (ANOVA) for the RCB design shown in Fig. 2.1 and for a completely randomized design using the same resources are shown in Table 2.1.

We have *paid* 5 d.f. to achieve the blocking shown in Fig. 2.1 and, if it happened that there was no greater variation between distant plots (blocks I and VI) than between adjacent ones, we would have *reduced* the precision of the experiment to the extent that there are fewer degrees of freedom for conducting '*F*' or '*t*' tests on treatment means. We shall see below that it is possible to calculate, *after completion of an experiment,* whether or not blocking has improved precision. Unfortunately, we cannot conduct this analysis *before* we do the experiment, but we can make informed judgments about the likely size of block effects.

In many animal experiments, the individual animal is a '*plot*' to which we will apply a treatment. In the rest of this chapter we shall refer to 'animals', meaning plots; but we should note that, in some experiments, the animal may be given a treatment for a limited period and then switched to another treatment so that the plot becomes an 'animal-period' (see Latin square and other change-over designs in Chapter 5) and that the plot is sometimes a *group* of animals sharing the same pen or paddock (see Chapter 6). In other cases, a single 'plot' may be a synthetic fibre bag to be suspended in the rumen of a sheep or a laboratory vessel used for an *in vitro* digestibility study.

Given a number of animals to be treated as individuals and a number of treatments, we have a choice of allocating treatments entirely at random or of grouping the animals into blocks. *Attributes* commonly used for blocking are liveweight, age, parity,[a] litter size (singles versus twins in growing lambs, or predicted litter size in pregnant ewes), previous yield of milk or eggs, and growth rate prior to the start of an experiment. Note that these are all things that

Table 2.1. Frameworks for the analysis of variance of an experiment with six replications of four treatments, with or without blocking.

Randomized complete block design		Completely randomized design	
Source of variation	d.f.	Source of variation	d.f.
Blocks	5		
Treatments	3	Treatments	3
Error	15	Error	20
Total	23	Total	23

[a] Parity has two meanings: in everyday English it means 'equality', but in medicine and agriculture it is used as a technical term to mean 'number of pregnancies'. Thus a third-parity sow is one carrying, or having delivered, her third litter.

can be measured in animals *before* any experimental treatments are applied. Allocating animals to blocks on the basis of measurements taken after the experiment has started is cheating (but see Chapter 10 on Covariance, which may be helpful). *Position* in the animal house (which might seem to be the direct analogy of Fig. 2.1) is usually *not* a good basis for blocking animal experiments (but see 'Confounding' below).

Faced with the questions 'Should we assign these animals to blocks?' and 'What is the best feature to use for blocking?', we need to ask another question: 'What is the expected *correlation* between the things we have measured as a possible basis for blocking and the traits that we plan to use as measures of response to treatments?' For example, if we know the starting liveweights of animals to be used in an experiment, we need an estimate of the correlation between starting weight and things to be measured under treatment, such as growth rate, feed intake or carcass composition. Now, we cannot, at the start of the experiment, estimate these correlations directly: we therefore have to rely on correlations measured in earlier trials, together with our general knowledge of how animals perform.

It may, at first sight, seem obvious that animals that are bigger at the start of a trial are likely to grow faster; but if the animals in question have recently been weaned, differences in their starting weights may largely reflect differences in dams' milk supply and, under subsequent common feeding conditions, the smaller animals may grow faster than the big ones. This does not matter for purposes of blocking – provided there is a correlation, the blocking will be useful, whether that correlation is positive or negative – but the example is given to illustrate the pitfalls of making 'obvious' assumptions. There may be *no correlation* at all between the character you have chosen as a basis for blocking and any of the traits that you measure subsequently. If that is the case, then you will have paid for blocking in degrees of freedom and obtained only disadvantage from the investment.

The question of what is, and what is not, likely to be a useful character for allocating animals to blocks is one that is best answered by the experienced *animal scientist*, not the consulting statistician, although the statistician can be very good at asking pertinent questions.

The '*price*' of blocking is much greater in small trials than in large ones. If we have 24 animals to allocate to four treatments, then the error d.f. will be 15 with blocking and 20 without (see Table 2.1). If we suppose that blocking is totally ineffective (i.e. the blocks mean square (m.s.) equals the error m.s.), the loss of precision is then related to Student's t values (see Appendix 26) for 15 and 20 d.f. (2.131 and 2.086 for $P = 0.05$). The trial with blocking would give a precision 2.086/2.131 = 0.979 of that obtainable without blocking – a loss of only 2.1% in precision. On the other hand, if there were only six animals to be allocated to two treatments, the corresponding error d.f. are 2 (with blocks) and 4 (without blocking). Inspection of the t values for 2 and 4 d.f. (at $P = 0.05$) shows that blocking will lead to a 35% *loss of precision* in this case, *unless the blocking is effective* (i.e. unless the blocks m.s. turns out to be greater than the error m.s.). Thus, where you have plenty of resources, you might follow a policy of blocking every time, on the basis that the cost is small and there will often be a worthwhile reduction in the error variance (variance and m.s. are the same thing, in case you had forgotten). But, with small trials yielding few d.f. for error, the price of blocking may be too high, and it is worth making diligent enquiries about the likelihood that blocking will isolate a substantial portion of variation between animals. A further example of the price of blocking being too high is given under the discussion of Latin square designs in Chapter 5.

At this stage it might be useful to turn to Exercise 2.1 at the end of this chapter to remind yourself how to carry out an ANOVA for an RCB design.

Allocating Animals to Blocks and Treatments

This is a simple matter but does sometimes cause confusion. Suppose that we have 20 goats to be used in an experiment comparing four treatments and that we decide to allocate them to blocks on the basis of liveweights measured a few days before the experiment is due to start. Table 2.2 gives a list of the goats and their weights.

The first step is to rearrange the list so that animals are ranked in order of weight (see below). As there are to be four treatments, we then divide the list into five blocks of four, such that animals within the same block are as similar as possible and the blocks are as divergent as possible.

The second step is to allocate animals to treatments within blocks strictly at random. This is best done by using tables of random numbers as described in Appendix 1. An alternative (for four treatments) is to roll an unbiased dice, ignoring numbers 5 and 6 when they come up. What you must *not* do is to pay any attention to the outcome in block I when you come to allocate treatments in block II. There is a temptation to say that, because the heaviest animal in block I happened to be assigned to treatment C, we must then make sure that treatment

Table 2.2. Initial liveweights of goats to be allocated to blocks and treatments for an experiment.

Goat no.	Weight (kg)	Goat no.	Weight (kg)
1	25	11	27
2	32	12	39
3	38	13	34
4	30	14	23
5	22	15	33
6	28	16	41
7	35	17	30
8	33	18	27
9	27	19	28
10	29	20	38

C goes to one of the lighter animals in block II 'to even things up'. This is *cheating* and has serious consequences. It moves variation out of the treatment m.s. and into the error m.s. An analysis of variance of the preliminary data used for blocking should show similar (or not significantly different) variances for treatments and error (and a much larger variance for blocks), although you will understand that once in every 20 trials you can expect a preliminary analysis to show 'significant' differences between 'treatments' (which have not yet been applied) and error purely by chance. What you should do if you find that there is a significantly large 'treatment' effect at the outset is a matter of philosophy. There is much to be said for the philosophy of the late, great H.G. Sanders who advised that 'there can be no justification for altering a random allocation once it has been honestly carried out; but anyone can be forgiven for losing a piece of paper!' Our piece of paper, after the allocation to blocks and treatments might look like Table 2.3. An ANOVA of these preliminary weights is shown in Table 2.4.

The blocks m.s. is significantly large, of course, because we have made it so. The treatments m.s. is *curiously small*, leading to a suspicion that the experimenter fiddled the random allocation; but the F-ratio (2.358/0.317 = 7.44 with 12 and 3 d.f.) tells us that such a close correspondence of initial treatment

Table 2.3. Listing of goats ranked in order of initial weight and allocated to five blocks on the basis of weight and to four treatments at random within blocks.

Rank	Weight (kg)	Goat no.	Block	Treatment
1	22	5	I	B
2	23	14		D
3	25	1		A
4	27	9		C
5	27	11	II	A
6	27	18		D
7	28	6		B
8	28	19		C
9	29	10	III	B
10	30	4		C
11	30	17		A
12	32	2		D
13	33	8	IV	C
14	33	15		D
15	34	13		A
16	35	7		B
17	38	3	V	C
18	38	20		D
19	39	12		A
20	41	16		B

Table 2.4. Analysis of preliminary weights of goats allocated to blocks and treatments as in Table 2.3.

Source	d.f.	s.s.	m.s.	F
Blocks	4	519.70	129.925	55.1***
Prospective treatments	3	0.95	0.317	
Error	12	28.30	2.358	

means would be expected by chance in something between 1 in 10 and 1 in 20 trials (F values are tabulated in Appendix 25).

Improvement in Precision Due to Blocking

We can use the results of Exercise 2.1 to illustrate how to calculate the effectiveness of blocking *in retrospect*. The ANOVA of carcass weight from that exercise is given in Table 2.5.

Blocks have accounted for a significant chunk of variation in carcass weight and so it was a good idea to block the animals on initial liveweight in this case.

To work out exactly what has been gained by blocking, we first calculate that, without blocking, the error m.s. would have been:

$$\frac{29.1322 + 19.3616}{7 + 21} = 1.7319 \text{ (28 d.f.)}.$$

The error in the RCB design is only (0.92198/1.7319) = 53% of what it would have been without blocking. However, we must also allow for the cost in lost d.f. For a test of treatment differences we have 3 and 21 d.f. in the RCB design, for which the F value at $P = 0.05$ is 3.07, but 3 and 28 d.f. for the completely randomized design with an F value of 2.95. The efficiency of the RCB design relative to complete randomization in this example is therefore:

Table 2.5. Analysis of variance of carcass weights listed in Exercise 2.1.[a]

Source	d.f.	s.s.	m.s.	F
Blocks	7	29.1322	4.1617	4.51**
Diets	3	10.4359	3.4786	3.77*
Error	21	19.3616	0.92198	
Total	31	58.9297		

[a] This table, in common with others throughout the book, presents more decimal places than would be appropriate in a published paper. This has been done deliberately so that students who perform their own calculations (as strongly recommended by the author) can verify that they have obtained correct solutions. All these decimal places should be carried during the calculation, but resulting mean squares can be rounded to three significant figures for publication.

$$\frac{1.7319 \times 2.95}{0.92198 \times 3.07} = 1.81.$$

Thus the RCB design was 81% more efficient than the completely randomized design in this case. If blocking had not been used, treatment effects would have needed to be nearly twice as large to be detected as significant.

Double Blocking

There is no reason (except the price in d.f.) why *two or more* features should not be used as restrictions on the random allocation of treatments, with consequent adjustment of the ANOVA.

For example, we may have *male and female* lambs to use in an experiment and also know the starting weights, which we believe will correlate with feed intake and growth rate while on the experiment. In this case, the obvious procedure is to allocate lambs to blocks (defined by starting weight) *within sexes*. All the males are ranked in order of initial weight and divided into blocks and then the procedure is repeated with the females.

Exercise 2.2 at the end of this chapter gives an example of this kind of double blocking. A point to notice, is that it is easy to get an *inappropriate ANOVA* in this case if sex is labelled as a 'treatment' which, in a sense, it is, and then we go on to say that we are using a factorial design in which 32 sheep are arranged in four blocks with 2 × 4 = 8 (sex × diet) treatments allocated within blocks. This description would involve grouping together 'ewe block I' and 'wether block I', 'ewe block II' and 'wether block II', etc. and would lead to ANOVA A in Table 2.6.

There are fewer d.f. for error in ANOVA B, which might seem a disadvantage, but what has been isolated (in the 'Blocks' term up above) is a component with 3 d.f. representing the extent to which the *difference* in carcass weight between ewes and wethers varies between blocks I, II, III and IV. This might be a substantial component (if the range of weights happens to be different for ewes and for wethers) and it is worth paying 3 d.f. out of 21 to remove from the error term.

Another common example of double blocking arises where two or more *breeds* are involved in an experiment although, again, 'breed' may be treated as a block effect or it may be allocated as a treatment. The latter is more appropriate if it is important to estimate the breed difference reliably, as opposed to simply removing it from error. Figure 2.2 shows three possible layouts for a chicken experiment carried out in laying cages, in which three diets are to be tested on two breeds of chicken.

In design A, 'breeds' represents a grouping of the blocks. This gives maximum precision for testing the average effects of diet (because the blocks are smaller) but will not give a dependable estimate of the mean difference between breeds. In design B, breeds are regarded as 'treatments', resulting in a

Table 2.6. Analyses of variance of carcass weights from Exercise 2.2 regarding sex as a treatment (ANOVA A) or treating sex as a component of blocking (ANOVA B).

ANOVA A Treated as 4 blocks with 2 × 4 treatments/block			ANOVA B Treated as 8 blocks (4 ewe blocks + 4 wether blocks) with 4 treatments/block		
Source	d.f.		Source	d.f.	
Blocks	3		Blocks	7	
Treatments	7		Sexes		1
Sexes		1[a]	Blocks within sex		6
Diets		3	Diets	3	
Sex × diet		3	Blocks × diets	21	
			Sex × diet		3
Error	21		Error		18
Total	31		Total	31	

[a] Where d.f. are offset to the right, as in these two analyses, this is to indicate that they represent a subdivision of a previous component and so must not be added in when reckoning the error d.f. by difference. Both of these ANOVAs are basically RCB designs; sex is a subdivision of the treatment s.s. in A but a subdivision of the blocks s.s. in B.

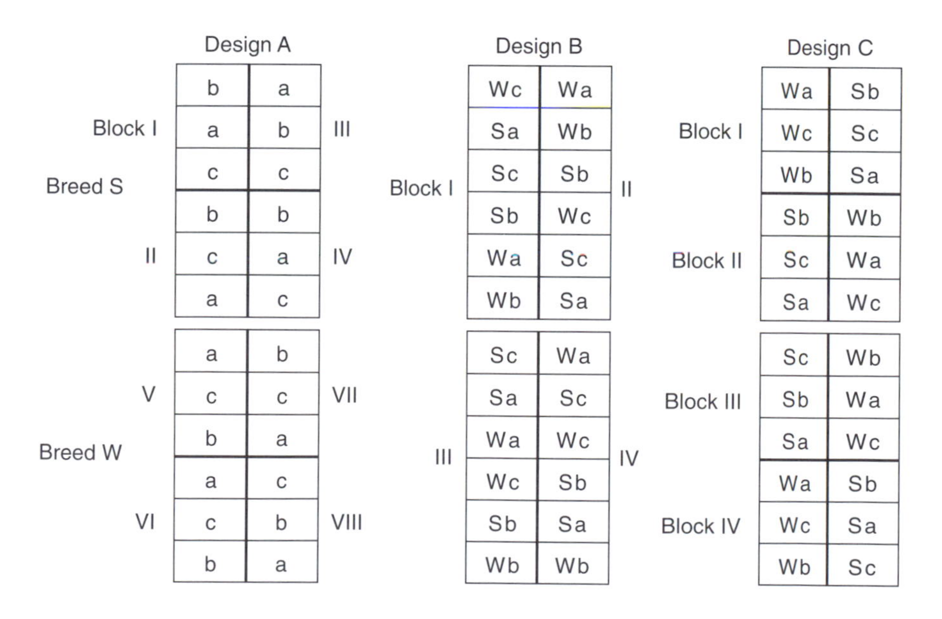

Fig. 2.2. Plans of three possible designs for an experiment with laying hens in cages, testing the effect of three diets (a, b and c) on two breeds. Each plot represents a *group* of chickens.

factorial design (the factors investigated being 'breeds' and 'diets') with 3 × 2 = 6 treatments arranged in four randomized blocks (the analysis of factorial experiments is discussed in Chapter 7). Design B will give a much better estimate of the difference between the breeds *and* of the breed × diet interaction, which may be just as important to the experimenter as the main effect of diet.

A third possibility is to allocate the breeds randomly within four (larger) blocks and then to allocate the diets independently within these main plots (design C). This is a split-plot design (see Chapter 7) which will have two errors, one for testing breed effects and another (probably smaller) error for testing treatment differences and the breed × treatment interaction. Design C sacrifices some precision in testing breed differences (compared with design B) but is likely to give greater precision for the comparison of diets (it is equal to design A in this respect). Design C is, of course, better than design A for testing the effect of breed and the breed × diet interaction, and so the only good reason for choosing design A would be an operational one (it might be easier to manage the flock if breed S is in one bank of cages and breed W in another). Table 2.7 gives the correct ANOVA structures for the three designs illustrated in Fig. 2.2.

Another, more straightforward, example of double blocking is the case of dairy cows which may be of different ages or (more usefully) parities and for

Table 2.7. Frameworks for the analysis of variance for the three designs illustrated in Fig. 2.2.

ANOVA A		ANOVA B	
Source	d.f.	Source	d.f.
Breeds	1	Blocks	3
Blocks within breeds	6	Breeds	1
Diets	2	Diets	2
Breed × diet	2	Breed × diet	2
Error	12	Error	15
Total	23	Total	23

ANOVA C

Source	d.f.
Blocks	3
Breeds	1
Error (a)	3
Main plots	7
Diets	2
Breed × diet	2
Error (b)	12
Sub-plots	23

which a second sensible basis of blocking is the order in which they calve. Thus we might first classify the available animals into **(i)** heifers; **(ii)** second-calvers; **(iii)** older cows; and then make up blocks within each category. With three treatments to allocate, the first three heifers to calve become block I, the next three block II, and so on. There is no requirement to have the same number of blocks in each category. We might, for example, find ourselves with:

9 heifers = 3 blocks
6 second-calvers = 2 blocks
12 older cows = 4 blocks
total = 27 animals in nine blocks of three.

The ANOVA for this grouping is shown in Table 2.8.

The same comment about unequal numbers of blocks within categories applies to other kinds of double blocking (e.g. breeds or sexes). The only requirement, for an RCB design, is that the number of plots in each block should equal the number of treatments. There are ways of analysing experiments in which this condition is not satisfied, but these are usually non-orthogonal designs requiring special analytical methods, as discussed in Chapter 11.

Confounding

It is possible to allow for the effect of *position* in an animal house by making it the basis of blocking and sometimes this will lead to a standard RCB design. If another feature, such as initial weight, is also thought to be important, this may give rise to double blocking. However, an alternative which often meets the need is to *confound* the two features so that their effects can be isolated from the error, but not separately identified.

'Confounding' is a general term used by statisticians to mean that two effects are so mixed up that they cannot be separated. An extreme example would be a trial using two-year-old heifers and three-year-old steers. Suppose that we find there is a difference in fatness between these two groups, we have no way of telling whether this is due to age or due to sex because the two effects

Table 2.8. Analysis of variance framework for an experiment with dairy cows using blocks based on parity and calving date.

Source	d.f.
Parity	2
Blocks within parities	6
Treatments	2
Treatment × parity	4
Error	12
Total	26

are completely *confounded*. Deliberate confounding of effects when designing experiments is a skilled art which will often call for professional advice from a statistician; but the simple case where we wish to confound positional effects with blocks (determined on some other basis) is straightforward enough.

Suppose that goats allocated to blocks on initial liveweight (as in Table 2.3) are to be placed in 20 individual pens, we might choose a layout as shown in Fig. 2.3. Here the effect of position in the barn is completely confounded with blocks based on initial weight. The ANOVA will have the structure shown for analysis of preliminary weights in Table 2.4, but the term labelled 'blocks' will isolate variance due to initial weight *and* position. If we are not particularly interested in estimating the magnitude of these effects, but only in removing them from our measurement of error, then the design is efficient for our purposes. If, on the other hand, we do want to know whether initial liveweight itself is an important source of variance or whether there really are position effects in the barn that might influence the design of future experiments, then we shall need a design that uses double blocking rather than confounding. However, with only five replicates it is not easy to devise a design which, by double blocking, will satisfactorily estimate both position effects and initial weight effects, and this is one good reason why confounding is often a sensible practice.

Pen	Goat	Treatment	Block
1	10	B	
2	17	A	III
3	4	C	
4	2	D	
5	18	D	
6	11	A	II
7	6	B	
8	19	C	
9	3	C	
10	20	D	V
11	16	B	
12	12	A	

Pen	Goat	Treatment	Block
20	13	A	
19	8	C	IV
18	7	B	
17	15	D	
16	9	C	
15	1	A	I
14	14	D	
13	5	B	

Fig. 2.3. A layout for an experiment in which goats were blocked on liveweight and allocated to treatments within blocks at random (as in Table 2.3) and then allocated to individual pens so that weight blocks are confounded with position in the animal house.

Block × Treatment Interaction

In a standard RCB design the error term *is* the block × treatment interaction. That is, the *error* measures the extent to which *treatment differences vary from block to block*. But suppose that we block a set of animals on the basis of their initial liveweights and there is reason to suspect that treatment responses may be different in the larger and the smaller animals. With a standard RCB design, we would have no way of asking whether or not it is true that the big animals gave significantly greater (or smaller) responses than the small ones.

One simple device for getting round such a difficulty is to increase the number of plots allocated to each treatment within a block so that, instead of placing only one plot of each treatment in each block, we choose to have two or more, with a corresponding reduction in the number of blocks. Thus, if there were 24 animals to be distributed to four treatments we could choose a block size of eight and allocate *two animals to each treatment in each block*. The ANOVA would then be as shown in Table 2.9.

Another example can be drawn from the 20 goats listed in Table 2.3. These could be divided into two blocks, with the 12 goats weighing up to 32 kg assigned to block I and the eight goats over that weight assigned to block II. The ANOVA would then be as in Table 2.10.

With a range of weights such as given for the goats in Table 2.3 we may not be worried that there is a serious risk that 'big' goats and 'small' goats will give different treatment responses, particularly when we note that most of the goats in

Table 2.9. Analysis of variance framework for a design chosen to reveal block × treatment interaction, with 24 animals allocated to four treatments.

Source of variation	d.f.
Blocks	2
Treatments	3
Blocks × treatments	6
Error (differences between treatment effects within blocks)	12
Total	23

Table 2.10. Analysis of variance framework for a design using 20 animals in two blocks, with three animals on each of four treatments in block I and two animals on each treatment in block II.

Source	d.f.
Blocks (initial weight)	1
Treatments	3
Treatments × initial weight	3
Error	12
Total	19

both of these blocks are inside the range 31 ± 7 kg. But, in other cases, there may be a much wider range of sizes or ages amongst the animals available for a trial and using a conventional RCB design may then run a serious risk of throwing a genuine block × treatment interaction into the error term, from whence it cannot be recovered. The moral is that we should pause to ask ourselves, when assigning animals to blocks, whether we have any reason to *expect* a block × treatment interaction in that particular case.

You may like to note that covariance analysis (which is discussed in Chapter 10) is often a good way of making allowance for initial differences in weight (or any other character) but, because routine covariance analysis uses the average *within treatment* regression to make adjustments, it does nothing to test or reveal whether *treatment responses* are influenced by the trait used as a covariate. A discussion of the comparative merits of blocking and covariance will be found in Chapter 10.

Summary

1. The *purpose* of blocking is to increase precision.
2. The *cost* of blocking is a reduction in error d.f.
3. Blocking should only be done if there are grounds for thinking that the *loss* of error d.f. will be more than *outweighed* by a reduction in error variance.
4. Judging whether blocking will be effective depends on guessing the *correlation* between the character used for blocking and subsequent measurements of response to treatments. This guessing should be done by an experienced animal scientist, not a statistician.
5. It is possible to use *more than one criterion* as a basis for allocating animals to blocks (e.g. sexes *and* initial weight).
6. It is sometimes useful to *confound* position effects with some other feature used for blocking.
7. Some thought should be given to the possibility of *interaction* between blocks and treatments.

Exercise 2.1

Perform an analysis of variance of the data in Table 2.11, taken from an experiment in which four dietary treatments were compared, with eight sheep allocated to each treatment in a randomized complete block design. The blocking was based on liveweight of the sheep at the start of the trial.

A method for doing this exercise with a pocket calculator is given in Appendix 3. A method using the SAS computing package is given in Appendix 4.

A discussion of the *improvement in precision* resulting from blocking in this example will be found on p. 9, where the ANOVA is also revealed.

Table 2.11. Dressed carcass weight (kg).

Block no.	Basal diet	Basal + acetate	Basal + propionate	Basal + butyrate	Block totals
I	16.3	18.9	19.4	18.0	72.6
II	16.4	18.2	17.6	17.5	69.7
III	16.7	18.9	17.6	18.6	71.8
IV	17.7	19.5	19.8	19.1	76.1
V	18.0	17.4	19.3	18.4	73.1
VI	19.1	18.0	16.5	17.6	71.2
VII	19.1	21.0	18.9	21.3	80.3
VIII	18.0	21.3	19.9	21.1	80.3
Treatment totals	141.3	153.2	149.0	151.6	595.1

Exercise 2.2

The data in Table 2.12 are from an experiment in which growing sheep were fed a basal diet and the same diet plus isoenergetic amounts of salts of three fatty acids. Four ewe lambs and four wether lambs were started on each diet and the animals were allocated to blocks according to initial liveweight within sexes. The sheep were all penned separately and the allocation of sheep to pens was random.

Analyse the data and write a summary of the conclusions to be drawn. (If you have already completed Exercise 2.1, note that the numbers in Exercise 2.2 are identical. This will save you the trouble of repeating some calculations.)

A method for completing this exercise using a pocket calculator is given in Appendix 5. A method using the SAS computing package is given in Appendix 6.

Table 2.12. Dressed carcass weight (kg).

	Block no.	Basal diet	Basal + acetate	Basal + propionate	Basal + butyrate	Block totals
Ewes	I	16.3	18.9	19.4	18.0	72.6
	II	16.4	18.2	17.6	17.5	69.7
	III	16.7	18.9	17.6	18.6	71.8
	IV	17.7	19.5	19.8	19.1	76.1
Wethers	V	18.0	17.4	19.3	18.4	73.1
	VI	19.1	18.0	16.5	17.6	71.2
	VII	19.1	21.0	18.9	21.3	80.3
	VIII	18.0	21.3	19.9	21.1	80.3
Treatment totals		141.3	153.2	149.0	151.6	595.1
Treatment means		17.663	19.150	18.625	18.950	

The framework of the analysis is discussed on p. 10 (see ANOVA B, Table 2.6). The separation of treatment means is discussed in Chapter 3.

Chapter 3

Separating Treatment Means

At the end of Exercise 2.2 we found that there were significant differences amongst the treatments, but some doubt may remain as to which treatment differs from which other treatment. The means with their standard errors are reproduced in Table 3.1.

The difference between the acetate and propionate treatments is 0.53 kg, which is less than the LSD. Butyrate is intermediate and so we may conclude that there are no differences shown by this experiment between the three fatty acid salts.

If, however, we test the difference between the basal diet and each of the other treatments, we find that acetate and butyrate yielded significantly heavier carcasses than the control treatment, whereas propionate did not (although it is very close to the 5% probability boundary). This might tempt us to a conclusion that propionate is not as good as the other two salts. Such a conclusion would be contrary to the finding in the previous paragraph that there were no significant differences amongst the three fatty acid salts.

This example illustrates two problems that arise in conducting t-tests to separate treatment means. One problem is *statistical* and the other is *philosophical*. We will deal with the statistical problem first, but it is the philosophical problem which is really important and interesting.

Table 3.1. Mean carcass weights of lambs from Exercise 2.2.

	Basal diet	Basal + acetate	Basal + propionate	Basal + butyrate
Treatment means (kg)	17.66	19.15	18.62	18.95

Standard error of each mean (SEM)[a] = 0.331 kg
Standard error of difference between two means (SED) = 0.469 kg
Least significant difference between two means (LSD), $P = 0.05$, = 0.985 kg

[a] If you have forgotten the definitions and methods of calculating SEM, SED and LSD, you may find Appendix 2 useful.

Statistical Problems with Student's '*t*'-test

You have probably met the warning that Student's *t*-test is only strictly valid for comparing *two* sample means. When there are, for example, four treatments and you compare all possible pairs, there will be six tests to conduct (AB, AC, AD, BC, BD, CD) and the probability of the means differing by chance is no longer 1 in 20, but some value greater than this. The point is easily illustrated by considering an experiment with eight treatments. There will be 28 *pairs* of treatments (7 + 6 + 5 ... + 1) and, even if the treatment means were derived by drawing numbers at random from a normal distribution, we should expect that the extreme high and low values would be judged 'significantly different', although the differences were actually due to chance.

The problem is not a serious one in trials with modest numbers of treatments. If you compare all possible pairs from a set of three or four means, the test is not strictly at the 5% probability boundary, but it is not far from that boundary and, if the observed differences comfortably exceed the LSD, there will be no need to look for more sophisticated tests.

Where the number of means is large, or when it is necessary to provide a more accurate assessment of probability, the solution usually offered is some form of *multiple range test*, such as the one due to Duncan (1955). These tests are sometimes built into computer packages and they are frequently used in reports in the literature, with superscripts against the means indicating which value differs significantly from which other value. There is nothing wrong with the statistical theory underlying multiple range tests, but there is a serious problem with the philosophy and you are strongly advised to avoid range testing in almost all circumstances.

The Philosophy of Separating Treatment Means

The procedure of putting treatments in a rank order and testing differences between them, either with Student's *t*-test or by means of a multiple range test, is very liable to lead to bad conclusions. The reason is that treatments almost always pose questions which are more subtle than the simplistic one: 'What is the order of merit of these treatments?' In some cases we set out to compare responses to a number of graded doses and this calls for a regression analysis, which is discussed under the heading '*dose–response trials*', below. In other cases we may be comparing a number of new treatments with a control that represents existing practice. Almost always, our choice of treatments can be represented by a set of logical questions and the analysis which we subsequently conduct should be related to those questions, or to a set of alternative questions formulated when we have seen the results.

Turning to the treatments in Table 3.1, there are three fairly obvious questions, which were presumably in the mind of the experimenter who designed the trial:

1. Does supplementing the diet with volatile fatty acid (VFA) salts cause an increase in carcass weight?
2. Are there differences amongst the VFAs?
3. If so, what is the relative fattening potential of these three VFAs?

We may note in passing that the experimenter almost certainly believed he knew that the answer to question 1 would be 'yes'. It was necessary to include the basal treatment to measure the responses to supplementary VFA, but the experimenter would have been amazed if supplying additional energy in the form of a VFA salt did not lead to additional stored energy in the lamb carcasses. The experiment was not *designed* to answer question 1, but it is, none the less, an important question within the context of the experiment.

Question 1 calls for a comparison between the basal diet and the *mean* of the three supplemented diets. This can be done easily, either as an *F*-test within the ANOVA or as a *t*-test subsequently. If you need reminding how to compare one treatment with the average of three others, you will find a convenient crib sheet in Appendix 7. Being a comparison of just two things (basal versus the rest) there is 1 d.f. for this test and the *t*-test and *F*-test lead to *identical estimates of probability*. Question 2 is a comparison amongst three means: here there are 2 d.f. and an *F*-test does not ask the same question as a set of *t*-tests. The *F*-test is preferable and the method of calculation is shown in Appendix 7. The resulting ANOVA is shown in Table 3.2.

Note that the two component sums of squares (9.3126 and 1.1233) add up to the treatment s.s. If they do not, there is a mistake in the arithmetic (if you used a calculator) or in the methodology (if you used a computer). This also tells you that the s.s. 'amongst VFA salts' can be obtained by difference; but it is better to estimate it directly from treatment totals (Appendix 7), since that provides a useful check on all the calculations.

The analysis in Table 3.2 tells us that supplementation of the diet with VFA salts did yield significantly heavier carcass weights, but there was no indication

Table 3.2. Subdivision of the treatment sum of squares for the experiment listed in Exercise 2.2.

Source	d.f.		s.s.		m.s.	*F*
Sex	1		6.7528		6.7528	7.68*
Treatments (= diets)	3		10.4359		3.4786	3.95*
Basal versus rest		1		9.3126	9.3126	10.59**
Amongst VFA salts		2		1.1233	0.5617	–[a]
Sex × diet	3		3.5384		1.179	1.34
Error	18		15.8232		0.87907	

[a] There is no *F*-ratio in this box because the definition of an *F* value is that it is the greater of two mean squares divided by the lesser: thus *F* cannot be less than 1, although you may see such values in computer print-outs. We could ask, if we wished, whether the error m.s. is significantly larger than the 'amongst VFA salts' m.s.: that test would have 18 and 2 d.f., not 2 and 18.

of differences between the three VFAs. We can now draw up a summary of our conclusions from this analysis, which you may find interesting to compare with the summary that you made at the end of Exercise 2.2 (that is, if you did Exercise 2.2). Without using any statistical terms, we can say that:

1. Wether lambs yielded heavier carcasses than ewe lambs (19.1 versus 18.1 kg) but there were essentially no differences in the responses of the two sexes to dietary treatments.
2. Supplementing the basal diet with VFA salts increased carcass weight by an average of 1.24 ± 0.383 kg. There was no indication of significant differences amongst the acetate-, propionate- and butyrate-supplemented diets.

Note that this summary gives *numbers*, not just qualitative statements, and for the important response (VFA *versus* basal) it gives a standard error too. Try to avoid writing summaries that merely say that something was bigger or better or 'significantly different'. *It is the magnitude of treatment differences that is of primary importance, not their statistical significance.*[a]

Generalizing from the above example, we can draw up the following rules for making comparisons amongst treatment means that are *not* part of a dose–response series.

1. Ask whether there are any significant differences amongst the treatments, using an *F*-test.
2. Check whether there *might* be significant effects present, even if the overall *F*-test of 'Treatments' is not significant. With a large number of treatments, there may be significant differences between some subsets even though the *F*-ratio for 'Treatments' does not reach the 5% value. If you divide the treatment *sum* of squares by the error *mean* square this gives a number which is the *largest possible* *F*-ratio that might be obtained for a contrast involving 1 d.f. If this number exceeds the 5% probability *F* value for 1 and (error) d.f., it may be worth poking further.
3. If there are, or might be, significant differences amongst treatments, formulate a set of sensible hypotheses to be tested. These hypotheses should be set down as questions and the number of questions should *not exceed the number of degrees of freedom for treatments* (see the discussion of orthogonal polynomials in Appendix 7). It is also highly desirable that your hypotheses should form an orthogonal set (see Appendix 7 for a definition of *orthogonal*).
4. Produce an ANOVA with a component of variance representing each of the questions you have formulated.
5. Report your results as a table of means, with an appropriate SEM (sometimes more than one SEM will be needed if you have heterogeneous variances (see Chapter 8) or a split-plot design or unequal replication) and say how many d.f. are associated with each SEM (this will be the d.f. for the variance used in calculating the SEM).

[a] This sentence is a direct quotation from R.A. Fisher, the founding father of statistics as applied to experimental design.

6. Avoid, at all costs, a table cluttered up with superscripts purporting to indicate significant differences.
7. Discuss your results in the light of the hypotheses that you have formulated and tested.

An Appropriate Use of Multiple Range Testing

When, then, is it right to use Duncan's Multiple Range Test (or one of the other multiple range tests available)? The answer is: when you have a series of means with *no structure*. For example, if you determine the digestibility of 24 grass silages and 19 maize silages you may be interested in contrasting the mean values for the two types of silage but, within each type, and in the absence of information about harvest date or crop variety, you might well list the means in rank order and conduct a multiple range test (assuming that you have replicate values for each silage sample). This will determine which silage differs significantly from which other silage. However, you might think that the significance of differences in this example is not particularly important; it is probably more useful to talk about the mean, SEM and range of digestibility values for each of the categories of silage.

An example where testing the significance of differences amongst a set of ranked means might be more important is in the reporting of random sample tests of different stocks of chickens. If there are replicated entries for each stock, or if trials have been conducted at different locations, thus providing replication, it is possible to say, not only which company had the most profitable stock on sale, but also what confidence you have that the results demonstrate that the number 1 stock really is better than the number 2. This is a proper case for using Duncan's multiple range test.

The Meaning of 'Significance'

The interpretation of experimental results should not present any difficulty if all the treatments being compared are significantly different from each other at the 5% level of probability; but this is an uncommon result. Much more often some, or perhaps all, of the differences are not significant, yet you remain suspicious that the effects observed are real, although not reaching the conventional level of significance. You are right to be suspicious. The most common mistake made by beginners (and some who should know better) is to report that a treatment has had 'no effect' when, in fact, the treatment has produced the *effect that might have been anticipated*, although the experimental variation is such that one cannot confidently claim the observed difference to be the result of the treatments applied.

For the remainder of this discussion we will focus attention on the comparison of just *two* treatments. Consideration of a set of treatments and, in particular, a series involving incremental doses, will be taken up later in this chapter.

The process of testing the difference between two treatments with a 5% probability 't' value is very like a *criminal trial*. The jury at a trial has to decide whether a case has been proved 'beyond reasonable doubt'. Ordinary experience leads us to believe the following.

1. Guilty people are sometimes found 'not guilty', because the evidence against them is insufficient.
2. Occasionally, innocent people are convicted.

The procedure at criminal trials is designed to reduce the risk that innocent people are wrongly convicted (though mistakes do sometimes occur). Similarly, the statistical testing of experimental results is intended to *reduce the risk that false claims are made* for new discoveries, techniques or products, although mistakes will sometimes occur. One important difference between the scientific experiment and the criminal trial is that the experiment can be repeated an indefinite number of times and by independent investigators, thus creating new evidence and making it extremely unlikely that a new and surprising result will stand for long if, in truth, it is one of those 1 in 20 rogue trials that produce a 'significant' result by chance.

As experimental scientists, we are also expected to take into consideration evidence external to the experiment, which, in a criminal trial, would usually be categorized as 'inadmissible' (e.g. previous convictions or even previous unsuccessful prosecutions).

Suppose that there are several published experiments which show that the addition of an antibiotic to a chick diet causes an increase in early growth rate averaging about 5%, and suppose that we conduct a chick growth trial with a control diet and the same diet supplemented with an antibiotic. Say that the difference observed is 4% (in favour of the supplemented diet) and the LSD is 6%. We have failed to demonstrate, beyond reasonable doubt, that in this trial the antibiotic increased growth rate: but we have certainly not produced evidence that the antibiotic failed to do the job expected of it. A 4% response is not significantly different from zero, but it is also not significantly different from the 5% response which we had been expecting. If the antibiotic is a new one which we are contemplating marketing, we shall, of course, need more trials and stronger evidence before we can claim that this particular antibiotic does stimulate growth. If, however, the antibiotic is one that has been tested and shown to be effective in the past, we can conclude legitimately that it has stimulated growth by 4% (±6%) in this trial and add this result to other results already compiled to give an estimate of the true mean response to this antibiotic.

When results point in the direction anticipated, on the basis of either theory or previous evidence, we should be very cautious about the reporting of 'non-significant' differences. A common mistake is to say that the difference between a treatment and its control was not significant and, *therefore*, the treatment did not work in this case. This is a *perverse* interpretation of the data and represents a misuse of statistical testing. It is more informative to say that the treatment difference was not significantly different from zero but also not significantly different from the response expected on the basis of some reasonable hypothesis. Such an experiment, taken alone, has not advanced our understanding of the issue being investigated; but, taken in conjunction with other evidence, the trial may help materially in establishing the magnitude of the true response. It is possible, though unlikely, that results from a series of trials, *none* of which showed a significant difference between a pair of treatments, can be combined to demonstrate a statistically significant response. In real life, the more likely situation is that a series of trials, some of which showed a significant response whereas others did not, can be combined to show that the real difference between a pair of treatments is not only highly significant but also can be quantified with an acceptably small confidence interval. Some further commentary on combining the results of a series of trials will be found in Chapter 14.

Proving Two Things Equal

Sometimes the object of an experiment is not to test whether a new treatment can improve performance compared with a control, but to test whether a treatment is *as good as* the control. A commonplace example would be the substitution of an unfamiliar feedstuff (e.g. ground-up leather boots: crude protein content 400 g kg^{-1}) for a well-known feedstuff (e.g. soybean meal: crude protein content 400 g kg^{-1}). Diets containing the two raw materials can be fed to groups of animals and the result may be that the animals given ground bootleather grow at a slightly slower rate than the ones fed soybean meal, but the difference is not significant. The classical error in this case is to report that boot protein is just as good as soya protein because the difference in performance is not significantly different from zero. A more sensible approach would be to say that old boots are cheaper (even after allowing for the cost of grinding) than soybean meal and to calculate whether the bootleather diet would be more profitable than the soybean diet *on the assumption* that the true loss of performance to be expected when using bootleather protein is equal to the depression observed in this trial. You can, of course, put a confidence interval on your calculation of the change in profit resulting from the substitution of bootleather for soybeans, using the standard error of the treatment difference. That confidence interval may or may not embrace zero, but that is scarcely relevant since your *best estimate* of the change in profitability depends only on the mean difference. However, if bootleather turns out to be the cheapest option (according to your data), you would be wise to consider the confidence interval on that estimate before deciding to recommend a change of practice from using soybeans to using bootleather.

To prove that two treatments are *exactly equal* is logically impossible when using a finite amount of biologically variable material. The best that you can do, by using more and more replication, is to reduce the confidence interval between two treatments being compared until it is so small that no one will quarrel with your claim that the treatments really are equal. This is hard (and expensive) to do and, if you are dealing with large farm animals, you would be well advised to avoid, so far as possible, problems that require a demonstration that treatment A = treatment B.

The Balance of Probability Argument

There are some cases where the significance of a difference is irrelevant to the decision that has to be made. This is rather akin to the branch of civil law where a judge has to decide an issue, not as proven beyond reasonable doubt, but *on the balance of probabilities*. If the judgment has to be made either in favour of A or in favour of B, and the evidence suggests that there is a 55% probability that A is in the right (and only a 45% probability that B is right), then judgment will be given in favour of A. (Solomon found a trick for getting round this by ordering that the baby be chopped in half: but bluff does not always work that well!)

An example where a 'balance of probability' approach would be sensible would be the testing of two anthelmintics in sheep, as in Table 3.3.

The difference between the two products, A and B, is not significant. If B is a new product that you are marketing in competition with A, which has been around for some time, you would be entitled to say that the evidence indicates that your anthelmintic is just as good as A's (and that both are highly effective in improving lamb growth). However, if you are a farmer, and *supposing the two products to be equally expensive*, you would be sensible to choose A on the grounds that there is *some* probability that it is better than B. If B is cheaper than A, you can do a cost:benefit analysis and decide, on a balance of probability, which is the more cost-effective anthelmintic. The odds in favour of you having chosen the better of the two anthelmintics are not very high, but they are a little better than 50:50 and that is all you need to make a decision in this case. We should also note that the consequences of being wrong are not at all serious.

The morals of this section are:

1. Non-significant differences often convey useful information.
2. Do not assume that treatments shown to be not significantly different are necessarily equal.

You will have trouble with referees when you come to apply these precepts in your published papers; but careful wording of your conclusions can usually convey a sensible interpretation without offending even a pedant.

Table 3.3. Results of a trial comparing two anthelmintics administered to groups of ewes and lambs at pasture.

	Control (no worming)	Anthelmintic A	Anthelmintic B
Liveweight gain of lambs (g day^{-1})	103 ± 17	148 ± 9	134 ± 8

Dose–Response Trials

The most common error seen in published papers on agricultural science is the use of Duncan's multiple range test to interpret the results of a dose–response experiment. This is like trying to peel an apple with an axe (nothing wrong with apples, nothing wrong with axes, but they don't belong together).

As usual, we shall illustrate the mistake by taking a particular example. Suppose that an experimenter wishes to determine the protein requirement of growing pigs and that he compares four diets, with the results shown in Table 3.4.

The information in Table 3.4 is not false, but it is misleading. It may tempt the experimenter (and others) to draw the conclusion that the protein requirement for this class of pigs (using this quality of protein) is not more than 160 g kg^{-1}, on the grounds that the responses to higher levels than this were not significant. That is a *false conclusion*, not supported by the evidence. This is a case, as is true of almost all dose–response experiments, where testing the significance of differences between particular means is inappropriate and liable to lead to wrong conclusions. The difficulty stems from asking the question 'What is the requirement for protein?' instead of the much more appropriate question 'What is the response to protein?' As soon as the trial is recognized as an attempt to characterize the *response* of pigs (of this type) to protein (of this quality), we are led to seek an *equation* which will describe the response. A number of models might be used to interpret these data and a discussion of some of them can be found in a paper by Morris (1983) and in Chapter 9. The *simplest* alternatives (but not the best in many cases) are the straight line and the parabolic curve, which is described by a quadratic equation.

If we fit a parabolic curve to the data in Table 3.4, we get the result shown in Fig. 3.1. Also shown is the SEM for each treatment (which did not appear in Table 3.4, though it should have done). We can see immediately that the curve is a good representation of the data.

The equation of the curve is:

$$y \text{ (growth rate, g day}^{-1}) = -9283 + 119.5x - 0.345x^2$$

where x = dietary protein (g kg^{-1}). The corresponding ANOVA is shown in Table 3.5.

The treatment sum of squares in Table 3.5 has been divided by first fitting a linear regression, then adding a quadratic term and finally estimating a cubic component (all that is left over). You will notice that the linear regression accounts

Table 3.4. Growth rates of young pigs fed diets with four different concentrations of protein.

Protein in diet (g kg^{-1})	150	160	170	180
Liveweight gain (g day^{-1})	876[b]	1020[a]	1049[a]	1055[a]

Means with different superscripts are significantly different ($P \leq 0.05$).

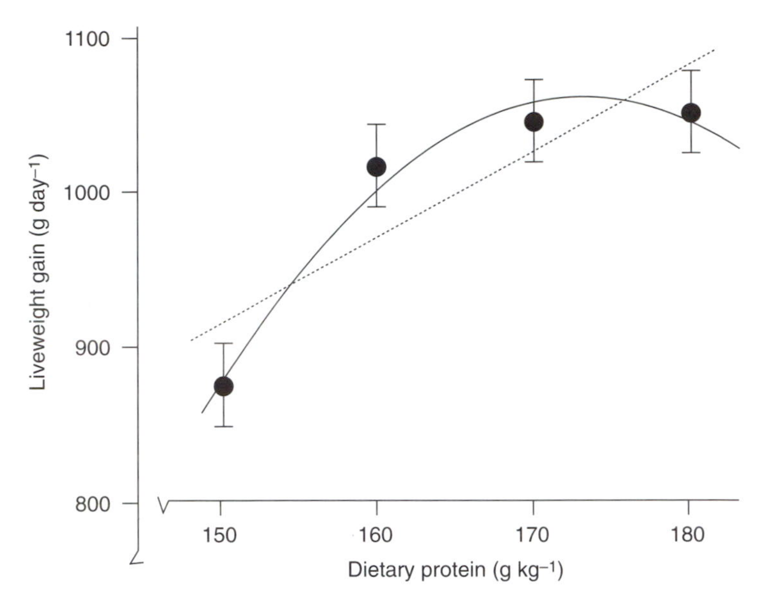

Fig. 3.1. Mean growth rate of young pigs given diets with four different concentrations of protein (n = 20 pigs per treatment).

for a highly significant chunk of the treatment variation but that the extra variation accounted for by adding a quadratic term is also significantly larger than the error.

There are asymptotic models, which might be more appropriate to this case than the quadratic model, and some of these are discussed in Chapter 9. However, with the data available (only four values and an SED that spans about 20% of the range) it is unlikely that we shall be able to show that another curve is a significantly better fit. This is not intended to deter you from fitting asymptotic models but only to point out that the choice of model has to be made on philosophical grounds, because one curve cannot usually be proven to be a statistically better fit than another.

Suppose that the experimenter accepts (grudgingly) the re-interpretation of the data so far (i.e. he accepts that a curve is a reasonable model) but persists with his original question 'What is the protein requirement?' That is still a bad question, but if some point estimate is demanded, we can offer at least two. We can estimate

Table 3.5. ANOVA for the pig experiment reported in Table 3.4.

Source	d.f.		s.s.		m.s.	F
Diets	3		424,040		141,346.67	10.43**
Linear regression		1		320,356	320,356.00	23.64***
Quadratic		1		95,220	95,220.00	7.03**
Cubic		1		8,464	8,464.0	
Error	76		1,029,721		13,548.96	

from the curve the protein level which we expect to *maximize* growth: this turns out to be[a] 173 g kg^{-1} which, in terms of dietary cost, is a materially different answer from the 160 g kg^{-1} implied by Table 3.4. Alternatively, we can estimate the protein level that will *optimize* growth, *provided that* we are given the *marginal* cost of increasing dietary protein and the *marginal* value of liveweight gain for pigs of this class. Neither of these inputs is particularly easy to define and, of course, both will vary with time and place. For these reasons, the objective of providing a point estimate of the optimum would be best discarded. The really useful thing to publish is the equation of the response curve, leaving others to decide, in the light of all sorts of local circumstances, how much protein they will put in their pig diets. All books entitled '*Nutrient Requirements of* …' should be renamed '*Nutrient Responses of* …' (with a corresponding adjustment of what appears between the covers of the book).

Summary

1. It is almost never appropriate to analyse the results of a formal experiment using a multiple range test.
2. If a trial measures responses to increasing doses of some input, look for an equation that sensibly represents the dose–response relationship.
3. If the trial does not involve different doses, look for a set of orthogonal questions (not exceeding the number of degrees of freedom for treatments) and test those questions in an analysis of variance.
4. Non-significant differences sometimes convey useful information, particularly when combined with information that is extraneous to the trial. Do not fall into the trap of thinking that two treatments have been proved equal just because they are not significantly different.

References

Duncan, D.B. (1955) Multiple range and multiple *F* tests. *Biometrics* **11**, 1–42.
Morris, T.R. (1983) The interpretation of response data from animal feeding trials. In: Haresign, W. (ed.) *Recent Advances in Animal Nutrition – 1983*. Butterworths, London, pp. 13–23.

Exercise 3.1

A chick trial was set up to find out which of several B vitamins might be responsible for an improvement in growth rate previously reported when

[a] If you have an equation of the form $y = a + bx + cx^2$, the maximum value of y corresponds to a value of $x = -b/2c$.

Magigro (a proprietary product derived from yeast) was added to an 'organic' diet (i.e. one not supplemented with synthetic vitamins). The results obtained are given in Table 3.6.

1. Suggest a subdivision of the treatment sum of squares which is likely to be helpful in the interpretation of these results.
2. Explain why, in an experiment of this kind, the strategy of omitting each vitamin in turn is likely to be more revealing than the device of adding each vitamin singly.

Table 3.6. Chick growth from 0 to 3 weeks on a diet containing no synthetic vitamins but supplemented with either Magigro or a mixture of five B vitamins, or the same mixture omitting each vitamin in turn.

	Growth rate (g day^{-1})
Organic diet	14.4
+ Magigro	18.3
+ mixture of five B vitamins	17.8
+ vitamin mixture omitting thiamine	19.1
+ vitamin mixture omitting riboflavin	13.8
+ vitamin mixture omitting niacin	16.2
+ vitamin mixture omitting pantothenic acid	18.6
+ vitamin mixture omitting pyridoxine	17.8
SEM	0.63

Exercise 3.2

Write down the questions you would ask of the data if you had obtained the experimental results given in Table 3.7.

Table 3.7. *In vitro* organic matter digestibility (OMD) of wheat straw treated with caustic soda at three concentrations or with aqueous ammonia at three concentrations.

	OMD (g kg^{-1})
Untreated straw	421
NaOH	
20 g kg^{-1} straw	480
40 g kg^{-1} straw	532
60 g kg^{-1} straw	599
NH_4OH	
20 g kg^{-1} straw	450
40 g kg^{-1} straw	497
60 g kg^{-1} straw	527
SEM	±31

Chapter 4

How Many Animals?

The most important – and the most often neglected – step in designing any experiment is the calculation of the number of replicates needed to give a reliable outcome. This is particularly true of experiments with large farm animals, where the cost per replicate is high, both in terms of the capital required to purchase and house the animals and the labour needed to feed them individually and collect records. Experimenters (and especially graduate students) commonly find themselves in the situation where they have been allocated a specified number of animals for their research and told to get on with it. That is not good enough. The proper response to such maltreatment is discussed at the end of this chapter.

In the following discussion, we will talk about 'numbers of animals', on the assumption that the individual animal is the experimental unit. As already noted there are cases where an 'animal-period' may be the true experimental unit (see the discussion of Latin squares in Chapter 5) or where a group of animals in a pen or paddock all receive the same treatment and the pen thus becomes the 'plot' (see Chapter 6). The principles of calculating the number of replicates required are the same for those types of experiment, but more care is needed in finding an appropriate estimate of the variability of the experimental units.

Calculating how many animals are needed to demonstrate a significant difference between two treatments is a matter of inverting the by-now familiar formula for an LSD. The least significant difference between two treatments with equal replication and a common estimate of error is:

$$\mathrm{LSD} = t \cdot \sqrt{2} \cdot \sqrt{(s^2/n)}$$

where t = Student's t value for a chosen probability and d.f. appropriate to the error variance; s^2 = the error variance; n = the number of replicates of each treatment.

If we define d as the difference which we wish to detect as significant in a trial that we are planning, then we must find what value of n will result in d being just equal to the LSD when the trial is completed. That is, we set

$$d = \mathrm{LSD} = t \cdot \sqrt{2} \cdot \sqrt{(s^2/n)}$$

Squaring both sides of the equation gives:

$d^2 = t^2 \cdot 2 \cdot s^2/n$

and therefore $n = 2\, t^2\, s^2 \,/\, d^2$.

This is the most important formula in this book and you would do well to learn it by heart (or learn one of the derivative, simplified versions given below). As t is a nearly constant number for any chosen level of probability, the important *variables determining* the number of animals required are s, the variability of our experimental material, and d, the size of the difference that we wish to detect.

Now, before going on to explain how we can obtain estimates of s and d to put into our equation, we should note that these parameters must be in the same units. It will not do to have s measured in kg milk day^{-1} and d in pounds per lactation. This difficulty can easily be got around, however, by turning both statistics into percentages of the mean. You probably know that a coefficient of variation (CV), expressed as a percentage, is defined as:

CV = 100s/general mean value.

We can do the same with d:

d% = 100d/general mean value.

This gives another form of the important equation above:

$n = 2\, t^2\, (\text{CV})^2/d\%^2$.

This equation has no units, except percentages, and is therefore simpler to work with. Now we are ready to ask the important question: 'How do we obtain estimates of CV, d% and t for an experiment which has not yet started?'

Estimating the CV of Future Experimental Material

The following is a list, not necessarily complete or in order of importance, of paths that you might take towards choosing an expected CV for an experiment that you are planning.

1. Use the result of an unsuccessful trial just completed. Researchers commonly find themselves in a situation where they have a result that looks interesting, but is not statistically significant. This naturally leads to the question 'Supposing this to be a real difference, how many replicates would I need to show that it is significant?' One can then take both d% and the CV from the completed trial to answer that question.

2. Take an average of several trials. It is often the case that, in a particular laboratory or station, numerous trials have been conducted measuring the same traits in similar experimental animals (for example, postweaning lean-tissue gain in rats, or plasma corticosterone concentrations in ostriches). It would then be sensible to calculate the CV for all the readily available local trials and take an average. (An arithmetic mean will serve your purpose: strictly, you should square the CVs, multiply each one by its d.f., sum the products, divide this sum by the total d.f. to arrive at a weighted mean square of the CVs and then take the square root; but don't bother!) Your search through past records might also throw up interesting differences which could lead to hypotheses about why the CV was greater in some trials than in others. If you believe the hypotheses are reasonable, you can then take steps that may help to reduce the CV in your own next trial.

3. Go to the library. If the body of locally available expert knowledge about coefficients of variation is small or non-existent, you can consult published reports.[a] This may give you a much longer list of CVs from previous trials, which would seem an advantage, but you will know less about the particular circumstances in which these estimates of variation were obtained. Small differences in procedure can sometimes have a substantial influence on the variability of results.

4. Measure some animals. This is not very helpful advice if your experiment is concerned with traits, such as survival rate or litter size, which can only be measured over a lengthy interval; but sometimes records are available from previous generations of animals and these may provide the estimate of CV you are looking for. If, on the other hand, you are interested in the effect of growth hormone treatment on the ratio of back length to shank length (a trait that might take many months of treatment before the *response* could be measured), you could quickly measure that ratio in some live animals and so obtain a CV before starting the trial.

5. Use an analogy. You are blazing a new trail and nobody has worked with *your* trait previously. So if you want an estimate of the CV for plasma glucose in llamas, find values reported for sheep. Or, if you want to investigate the effects of certain treatments on the weight of thigh muscle as a proportion of the carcass in ducks (which has not been reported) find some data for the variability of thigh muscle as a proportion of the carcass in chickens (which has).

6. Make an intelligent guess. This is not so hard as you might imagine. Growth rate in all mammals tends to have a CV around 12%. Reproductive characters such as litter size and fertility rates are much more variable, often with CV in the 20–40% range. Characteristics such as milk yield and egg yield, measured over many weeks not days, have a CV of 20–25%. Linear measures such as length of tibia or incisors (adjusted for body weight) are much less variable; CV about 6%. You may recognize these generalizations as the inverse of what we would say about the heritabilities of these characters. That is because heritability is the ratio of genetic variance to phenotypic variance (and phenotypic variance = s^2 in our equation above). A trait with a high heritability must have a small environmental variance and vice versa. So, if you can guess at the heritability of a trait, you can guess at its CV.

[a] The CV itself is rarely reported, but if there is an SEM or an SED or an LSD attached to a table of mean values, and if the number of replicates contributing to each mean is clear, it is then not a difficult task to work out the CV for individual experimental units. Appendix 2 may be helpful here.

Another important point to think about is the effect of experimental protocol on expected CV. If you measure plasma glucose immediately before feeding each day, the values will be much less variable (though perhaps less interesting) than if you measure plasma glucose while food is being digested. If you feed pigs *ad libitum* their growth rates will be more variable than if all pigs are allocated the same daily ration, and there will be an even greater difference in the CV of carcass fatness.

Absence of a prior estimate of CV can never be a valid reason for not calculating the number of replicates required. You must *make up some number*, using the most reasonable arguments you can think of, and fit it into the equation. If you don't like the answer (because it demands more animals than you have available), you can then think about which number to change (your guess about CV or your plan for $d\%$), but you will have been warned that you are heading for a wasted experiment unless you change something.

Estimating the Difference to be Expected

'If I knew what difference to expect between the treatments I wouldn't need to do the experiment.' This seems a reasonable objection but, in fact, experimenters can nearly always give good estimates of the magnitude of responses that they are expecting or hoping for. Here is a list of some approaches that may be helpful but, as with the list for CV above, the ideas are not exhaustive nor mutually exclusive, nor necessarily in order of importance.

1. Use the result of an unsuccessful trial just completed. Suppose you have just done an experiment which showed a non-significant difference between treatments of 7%. It would be perfectly reasonable to take $d\% = 7$ as the basis for designing your next experiment.
2. Survey the literature. You might do better, if this problem has been worked on previously, to look at a series of experiments, either from your own laboratory or from work published by others. This will give a more general estimate of the average value of $d\%$ and may lead you to conclude that the response varies with circumstances that you can identify (bigger in winter than in summer, or greater in young calves than in more mature cattle). This can be very helpful as it may persuade you to modify the protocol for your own experiment. Of course, anyone embarking on a piece of research should have surveyed the relevant literature before designing the first experiment, but it is interesting that this good advice is so often neglected. Searching for material on which to base an estimate of $d\%$ is one way of obliging beginners to read the literature *before* they do their first experiment.
3. Use theory. This is one of the best ways of estimating d, but the necessary theory is not always available. In the experiment described in Exercises 2.1 and 2.2, lambs were fed known amounts of volatile fatty acid salts as supplements to

a basal diet. There is enough theory to estimate, on reasonable assumptions about efficiency, the additional carcass fat that could be expected from those supplements. Moreover, if there is a theory about the relative efficiency of conversion of acetate, propionate and butyrate to stored fat, it is possible to estimate the expected differences in carcass weight between the three supplemented groups. Such theoretical calculations should always be carried out if possible and the experiment can then be seen as a test of a theory, which is what all scientific experiments should be about.

4. Test claims made by other authors. Sometimes experiments are designed to test the reliability of claims already made in the literature. In such cases it is natural to set *d%* at the level of the claim already made. If you hope that the claim is true, you probably say that you are setting out to 'confirm' a result. If you believe that the claim is false, you may say that you are setting out to 'test' the previous result. In either case the important point to remember is that your experiment has to be *capable* of showing that the claim is *false*. When completed, your trial must have a small enough LSD to draw a conclusion one way or the other.

5. Ask 'What would be an economically worthwhile response?' It is often the case with animal research that there is an economic objective and it may make good sense to work out what would be a cost-effective response to some new treatment. The cost of including a drug in the diet of a farm animal can be estimated as current market price or, if not yet on the market, the sale price that would make it worthwhile to go through the steps necessary to put it on the market. This can lead to an estimate of the additional yield needed to justify use of the drug and so provide an estimate of *d%*.

6. Ask 'How big a response is needed before farmers will adopt the new technology?' This looks, at first sight, like the same question as above; but there is a difference between asking 'What would be economic?' and asking 'What might be adopted?' Suppose that you are testing a feed additive for the broiler industry which you believe will improve the efficiency of feed utilization by 2% and that the additive will increase the cost of feed by 1%. This is a worthwhile proposition for the chicken industry and, what is more, large companies can, and will, conduct their own trials, involving tens or hundreds of thousands of chickens to confirm that your estimated 2% response is achieved in practice. Therefore setting out to demonstrate that a 2% response is significant is quite realistic in the context of industrialized broiler production. But suppose that you have developed a new feed additive for ruminants. The additional cost per tonne of concentrate may be 5% which, since ruminant feed is mainly forage, increases total diet costs by about 2%. Suppose that the additive reduces overall feed consumption by 6%, without affecting yield. This additive is clearly economic, but farmers are unlikely to buy it because, although they can readily recognize the 5% extra cost of the concentrate, they cannot recognize the 6% saving in feed consumption under farm conditions. The benefits of research have to be observable by customers before they will be adopted in practice.

Applying the Equation

Having selected some values for the expected CV and $d\%$ in your next experiment, we can fit them into the equation

$$n = 2\, t^2\, (\text{CV})^2/d\%^2$$

once we have chosen an appropriate value for t. It is up to us to choose a probability, and $P = 0.05$ will do as well as any; but we also need to know how many d.f. there will be. Now the d.f. will depend upon the number of replicates as well as the experimental design. We can supply the information about the intended design, but the number of replicates is what we are trying to calculate! So, we have a circular problem and the formal method of solving this is by *iteration*. That is, you guess at some number for n, use this to find t, apply the equation to get a new value for n, use that to find a new value for t and so on until n is not changing by more than a trivial fraction. However, there is a quicker short cut.

If you look in Appendix 26 at the value of t for $P = 0.05$ you will find that it falls from 2.09 for 20 d.f down to 2.0 for 60 d.f. So, for most practical purposes, t (for $P = 0.05$) ≈ 2 and our equation can be *simplified* to

$$n \approx 8\, (\text{CV})^2/d\%^2.$$

Only if the d.f. for error turn out to be less than 20 is it worth bothering to go through the iterations to find a more exact value for n.

Some Examples

Suppose that we wish to design an experiment with dairy cows for which we think that $d\% = 5$ would be a sensible estimate of the difference we wish to detect as significant (at $P = 0.05$). A survey of literature tells us that for a 36-week trial with high-yielding cows we can expect a CV of 25%.[a] Using the approximate formula above, we calculate:

$$n \approx 8\, (25)^2/5^2 = 200.$$

That is, *if* your cows have a CV of 25% you will need 200 cows *on each treatment* to detect differences of 5% between treatments as significant at $P = 0.05$! This is a frightening conclusion, but it is inescapable. Possible ways of reducing the CV are discussed in Chapters 5 and 10. Change-over experiments, discussed in Chapter 5 will certainly give a much smaller CV but they have some serious disadvantages. The use of covariance (Chapter 10) has few disadvantages and may well halve the CV. You will notice that halving the CV would *quarter* the number of animals required.

[a] That means that with a herd average of 8000 litres per lactation you can expect a range of individual yields from about 4000 to 12,000 litres. You may think that is a very wide range, but inspection of almost any set of dairy records will show that it is not unrealistic.

We may be prepared (or obliged) to reconsider the choice of $d\%$ in the above example. We might do this (rationally) if we decide to amend the treatments so that we can expect bigger differences or (irrationally) by hoping that the differences will turn out to be larger than we previously expected.

Let us try again with CV = 12% (which is reasonable for lactation trials with covariance adjustments) and $d\% = 8$. Now we calculate:

$$n \approx 8\ (12)^2/8^2 = 18 \text{ cows per treatment.}$$

This is more realistic, although, even now, we should note that there are few stations in the world which can find 72 well-matched dairy cows for a four-treatment experiment.

Exercise 4.1 gives a few more examples for practice, including one where iteration is required to obtain a suitable t value for the solution.

What Are the Chances of Success?

If, in our next experiment, the CV is 12% and if we use 18 replicates per treatment, then the LSD will be 8%. That outcome is certain, but the statement contains two 'ifs'. Sampling variation in the CV is something we may choose to ignore but there is a serious question whether the observed difference (d) due to treatment in our next trial will be equal to, greater or smaller than the 'true' difference (i.e. the value derived from an infinitely large number of trials) which we can call δ. The observed value of d will vary from trial to trial and a little thought should lead you to the uncomfortable realization that the chance that d in your *next* experiment will be *equal to or smaller than* the average value, δ, is no better than 50:50. This suggests that we must somehow allow for sampling variation in the quantity $(d - \delta)$.

These deviations of observed values from the true mean value follow a t distribution and we can design a trial to give a specified probability that d will be equal to or smaller than our chosen value by inserting a second t value into the equation that we derived previously. Thus:

$$n = 2(t_1 + t_2)^2 \cdot s^2/d^2$$

where t_1 = Student's t (as before) with a probability chosen by the experimenter and d.f. of the error variance; and t_2 = Student's t with d.f. of the error variance and a probability of $2(1 - p)$, where p = the probability of success we choose to have.

This is getting a bit complicated, so let us begin by seeing what happens if we choose a 50% chance of success (that is of obtaining a CV in our *next* experiment equal to or smaller than the long-term average). If $p = 0.5$, then $2(1 - p) = 1.0$. The t value for a probability of 1 is zero and so the t_2 value disappears and we are left with the original formula. If, on the other hand, we wish to have a *90% chance of success*, then $p = 0.9$ and $2(1 - p) = 0.2$. The t_2 value (assuming plenty of d.f.) for $P = 0.2$ is about 1.3 and the calculation of the number of cows required (from the preceding section) will become:

$n \approx 2\,(2 + 1.3)^2 \cdot 12^2/8^2 = 49$ per treatment.

This is a rather dramatic increase from the 18 cows per treatment needed for a 50% chance of success and may serve as a warning not to hope for very high chances of success when designing lactation experiments (or perhaps to avoid working with dairy cows if at all possible!).

Tabulation of Number of Replicates

It may have occurred to you by now that for *any* design problem where the CV = 12% and $d\%$ = 8, the number of replicates required (given a 50:50 chance of success) will always be about 18 (using the approximate rule above; the exact number will depend upon the error d.f., which will in turn depend upon details of the design that have to be specified). It is therefore possible to construct a table showing the number of replicates needed for a range of values of CV and $d\%$ and probabilities of success. Such a tabulation is given in Table 4.1.

You may now think that it would have saved a lot of bother if this table had been presented at the beginning of this chapter, cutting out all the steps leading up to it. But perhaps, now that you have read and understood the theory and the associated considerations, you will be in a better position to use the table wisely. Certainly, you still need some understanding of how to select a suitable CV and to estimate $d\%$ before turning to Table 4.1.

Table 4.1. Number of replicates required (n) to detect a given difference between two treatment means (d, expressed as a percentage of the mean value for the trait measured) as significant at $P = 0.05$, with coefficients of variation (CV) varying from 6 to 25% and with probabilities of success of 50% or 90%.

	Probability of success											
	50%						90%					
CV (%) →	6	9	12	15	20	25	6	9	12	15	20	25
	Number of replicates needed for *each treatment*											
$d\%$ ↓												
2	71	158	277				192	426				
4	19	40	71	110	194	300	48	109	191	296		
6	9	19	32	49	88	135	22	48	86	133	236	
8	6	11	19	28	49	77	13	27	48	75	133	206
10	4	8	12	19	32	49	9	18	31	48	86	133
12	3	6	9	13	23	35	6	13	22	34	60	92
16	3	4	6	8	13	20	4	8	13	20	34	53
20	2	3	4	6	9	13	3	5	9	13	22	34

The table is calculated for two-tailed t-tests and assumes a randomized complete block design with four treatments: this gives $3(n-1)$ d.f. for error. If there are fewer than four treatments, *more* replicates will be needed.

Other tables, similar to Table 4.1, can be found in other texts. You may notice small differences if you compare the number of replicates recommended by different authors. These are due to differing assumptions made when constructing the table. To find the appropriate d.f., one must state the number of treatments in the experiment being planned and assume a design structure.

The numbers in the left-hand half of Table 4.1 should be regarded as the *minimum number* of replicates for any experiment. If these numbers are not available, the experiment should not proceed. The numbers in the right-hand half of the table can be regarded as the *prudent numbers* to use. If the experiment is liable to cause distress to the animals, it is preferable to use a number of replicates that makes it unlikely that the trial will need to be repeated and this argues for using the 90% probability estimates rather than the minimum numbers given by the 50% probability calculations on the left.

What to Do if There Are Not Enough Animals

It frequently happens that, after calculating the number of replicates required, the experimenter discovers there is no way of obtaining that number of animals. He (or she) is then in much the same position as a civil engineer who is asked to design a bridge to span a chasm. The engineer performs the calculations, using the best available values for the strengths of materials and the forces that the bridge will have to withstand.[a] This leads to an estimate of cost of, say, $100 million. The commissioning authority then says that only $50 million is available and the engineer must do the best he can with that sum. If the bridge is built for $50 million, to its original dimensions but with reduced materials, it will fall down. The design engineer deserves to be locked up and probably will be. In the same way, if you proceed with an experiment that you know is inadequately replicated you are wasting resources and will only produce an unpublishable result (or, worse still, a result that gets published but only adds to the confusion in the literature).

When you find that the number of animals available is much fewer than the number required, there are three possible avenues of escape:

1. You can search for a satisfactory way of *reducing the CV*. Selection of experimental material sometimes helps (although if you are selecting from *within* an already fixed number of animals this is *not* likely to help). You *may* be able to change the trait you planned to record for one that is less variable (e.g. measuring ovulation rate in sheep instead of litter size or measuring plasma LH instead of ovulation rate), but remember that this may give you results which are more significant but less interesting. Ultimately you will need to know how your treatment affects litter size but, in the short run, you may be

[a] You will recognize these inputs as the analogy of your CV and *d* values. The numbers are not absolutely certain, but you (or the engineer) can adopt sensible estimates and so complete the calculations.

able to screen a larger number of potential treatments by measuring plasma LH.

2. You can sometimes amend your treatments so as to expect *larger* d% *values*. For example, instead of treating straw with ammonia at temperatures of 15, 20 and 25°C, try 15, 25 and 35°C (always provided that changing the question in this way does not destroy the purpose of the experiment). It may also be worth reading Chapter 9 on the choice of treatments for dose–response experiments. Widening the range of treatments can often yield a better model, capable of predicting the rate of response over small ranges that cannot be studied directly with sufficient precision.

3. You can *go to the library* and improve your knowledge by reading. This is by far the most sensible thing to do. If everyone faced with the problem of trying to conduct animal experiments with insufficient resources would go to the library and *think* about the problem, a great deal more progress would be made. Coming up with a new hypothesis or a fresh interpretation of the literature is much more valuable than doing another experiment and infinitely more valuable than doing a worthless experiment. Moreover, whether you are a research student or a supervisor of research students, it is the new hypotheses which you put forward that are important, not the new experimental data.

Summary

1. It is a simple matter to calculate the *number of replicates* needed for an animal experiment. Unfortunately, the exercise is seldom carried out.

2. The number of animals required goes *up* with the *square* of the *variation* of the experimental material and comes *down* with the *square* of the *difference* that one hopes to detect.

3. The number of replicates required for common situations has been *tabulated*, making it unnecessary to perform calculations in most cases.

Exercise 4.1

An experimenter is planning a growth trial in which pigs will be fed individually. Six diets are to be tested. The mean CV of liveweight gain in pigs is assumed to be 12% and the experimenter wishes to detect differences $\geq 8\%$ as significant at $P = 0.05$. Use the approximate formula

$$n \approx 8(\mathrm{CV})^2/d\%^2$$

to calculate the *total* number of pigs required for this trial.

What chance does the experimenter have, using that number of animals, of achieving his objective of an LSD $\leq 8\%$ of mean growth rate?

Exercise 4.2

A digestibility trial is to be conducted in which sheep will each receive a particular roughage *ad libitum*, with a daily supplement of concentrated feed. After 3 weeks, each sheep will be changed to another of the roughages. The mean CV of dry matter digestibility, using *periods within sheep* as the experimental unit, is 6%. Calculate how many sheep-periods will be needed to assess the digestibility of the four roughages with an 80% chance of achieving an LSD (at 5% probability) $\leq$ 10%.

How many sheep-periods would be needed if the experimenter says that she does not expect large differences between the roughages and that the purpose of her trial is to estimate mean digestibility for each roughage with a SEM ≤ 2 percentage digestibility points (the anticipated mean dry matter digestibility being 50%)? She also says that a 50:50 chance of achieving the desired SEM is acceptable.

Exercise 4.3

Calculate an additional row for insertion in Table 4.1, giving numbers of replicates required for $d\% = 5$, with a probability of success of 50% (CV ranging from 6 to 25%). Use the exact equation, with iteration if necessary, to make the d.f. agree with n, assuming four treatments and an RCB design. Round up your answers to whole numbers if the fraction of a replicate is ≥ 0.1.

Chapter 5

Change-over Designs

Change-over experiments are those in which each animal receives two or more treatments in succession and thus, in a sense, each animal becomes a block.

With two treatments, which we can call control and treated, the principle is very simple. The available animals are divided at random into two sets, one-half serving as controls while the others are treated. After a suitable *period* for measuring the responses, treatment and control are switched and a *second period* of measurement follows.

After the switch (and possibly after the initial introduction of treatments) it may be advisable to *discard* data for a week or two, to allow time for the new treatment to take effect. If you are measuring some continuous output, such as egg yield or milk yield, the discard time can often be judged retrospectively after examining the data. The advice of an experienced animal scientist (*not* a statistician) should be sought before choosing a discard period, particularly for those cases (e.g. a digestibility trial) where the response cannot be observed immediately by daily measurement.

In a change-over trial each animal, as we have said, becomes a block (though not necessarily a complete block if there are many treatments to evaluate). However, even with few treatments, the change-over is not a simple randomized complete block design because another element – *time* – has now entered the design. If we allocate eight animals to a treatment and keep eight as controls and then, after a suitable period of measurement, switch and continue measurements on the reversed treatments for a second period, we shall have an ANOVA as shown in Table 5.1.

Notice that the experimental unit is no longer a single animal but is what we can call an 'animal-period'. This has a powerful effect on the precision of the experiment, since each animal acts as its own control and the variability in *average* performance level *amongst animals* no longer enters the error term. Variability amongst animals in the extent to which they *respond to treatment* (that is the animals × treatment interaction) makes up the error term, except that we can chip away 1 d.f. which represents the change in average performance (upwards or downwards) between period 1 and period 2. This is useful in

Table 5.1. Analysis of variance framework for a change-over experiment with 16 animals, two treatments (treated *versus* control) and two periods.

Source of variation	d.f.
Amongst animals	15
Periods	1
Treated versus control	1
Error	14
Total (animal-periods)	31

lactation or egg production trials, for example, where there is good reason to expect that all animals will tend to give lower yields in period 2 than in period 1.

Change-over trials, then, usually give much *greater precision* than experiments in which the whole animal is the experimental unit; but this gain in precision is bought at considerable (and sometimes unrecognized) cost. The difficulty is that we have made an implied assumption that the animal's response to treatment is independent of whatever previous treatment it may have received. This is often not the case. To take a far-fetched example, suppose that the treatment involves vaccination against a disease. If we begin by vaccinating half the animals and then, in period 1 challenge them with the relevant live virus and observe their responses, we shall expect to measure a large difference between treated and controls. But now it makes no sense to re-label the treated animals as controls for period 2, because they will still be at least partially immune as a result of vaccination in period 1, and the previous control animals, which were challenged with live virus in period 1, are almost certainly immune and so not suitable for evaluating the response to vaccination in period 2. This is a clear case where the effects of treatment in period 1 *carry over* and influence the responses measured in any subsequent period. Although this is an extreme example and no one would be foolish enough to use a change-over design for a vaccination trial, there are many other cases where experimenters have used change-over designs without giving due consideration to the effect of treatments applied in the first period on animal performance in the second period.

We will return to the discussion of carry-over effects at the end of this chapter and will list cases where they probably do and probably do not arise. Before doing that it will be useful to discuss some other designs that have been devised to measure or control for carry-over effects of treatments and, as a preliminary to this, we shall consider Latin squares.

Latin Squares

You have probably already met the idea of a Latin square (LS). The name comes from ancient Roman tessellated pavements which were often laid out using different coloured tesserae in rows and columns such that each colour occurred once only in each row and each column. Field trials using the same restriction on the allocation of treatments have the advantage that they measure (and so remove from the error term) variation in two directions at right angles and thus control for variation in *any* direction, provided that it is a uniform trend (see Fig. 5.1a).

It would be possible to lay out an animal experiment to control territorial variation in two directions, using a LS design, without involving any change of treatments. Thus pens in a barn might be arranged as in Fig. 5.1a; but this is not a common use of the LS in animal trials, because positional effects are not usually important sources of variation. Much more common is the LS in which animals and periods take the place of rows and columns, as shown in Fig. 5.1b.

To form a LS, the number of animals and the number of periods must equal the number of treatments and this generally restricts their use to cases where four or five treatments are to be compared. A 3 × 3 square has only 2 d.f. for error, which is seldom enough, although you can replicate the squares, using animals in sets of three, and so build up enough d.f. to give the precision required. Squares larger than 5 × 5 take so long to work through that they are usually impracticable.

The ANOVA for the experiment illustrated in Fig. 5.1b is set out in Table 5.2 and a numerical example of the analysis of a 4 × 4 LS is given in Exercise 5.1 at the end of this chapter.

The analysis itself should present no difficulty, but there are assumptions underlying this design which should give us serious concern. They are as follows:

1. We are assuming no carry-over effects from one period to another. This may be reasonable for some treatments, but is often not a safe assumption for nutritional effects (see the discussion of carry-over effects at the end of this chapter).

(a)

Columns → Rows ↓	i	ii	iii	iv
1	B	A	D	C
2	C	B	A	D
3	D	C	B	A
4	A	D	C	B

(b)

Cows → Periods ↓	Daisy	Ruby	Spot	Gert
1	C	B	D	A
2	A	D	C	B
3	B	C	A	D
4	D	A	B	C

Fig. 5.1. Latin squares. (a) A field trial with four treatments (A, B, C and D); (b) an animal experiment applying four treatments (A, B, C and D) in a randomized sequence to four cows, with the restriction that each treatment appears once in each period.

Table 5.2. Analysis of variance for the experiment illustrated in Fig. 5.2b.

Source of variation	d.f.
Cows	3
Periods	3
Treatments	3
Error	6
Cow-periods	15

2. If this experiment is to measure milk yield, we are assuming *parallel lactation curves*. To the extent that the chosen cows show different rates of decline in yield during the course of the experiment, this will form a cow × period interaction. That interaction cannot be estimated: it is mixed in partly with the treatment variance and partly with the error variance (in unknown proportions – it depends on the exact nature of the interaction). All this is bad news.
3. We are assuming that it is useful to measure the short-term yield response to treatment and that it is unnecessary, for purposes of this trial, to know how yield would be affected over a whole lactation (or a whole winter feeding period).
4. We are assuming that changes in liveweight reflecting changes in body condition are unimportant in relation to the questions we are asking. This is because the liveweight of a ruminant animal has a large error of estimate (due to variations in gut-fill) and it is not possible to make useful estimates of true changes in body weight in individual animals over intervals as short as 4, 5 or 6 weeks.

In relation to the first of these concerns, something can be done to estimate carry-over effects between one period and the next. By using a *balanced LS* design (see below), the effect of each treatment on performance in the immediately following period can be estimated. But before describing these designs there is a point worth making about degrees of freedom in the LS.

It is sometimes argued that paying d.f. to estimate between-animal and between-period variation is an unwarranted sacrifice of error d.f. in some types of experiment (e.g. digestibility trials with sheep). This is based on a finding that animal and period variances in such trials are often of the same magnitude as the error variance. In other words, digestibility does not vary systematically with time (not surprising) and there are no consistent differences between sheep (of the same age and type and receiving similar dry matter intakes) in their digestibility coefficients. Thus, it is argued, one might as well use a design such as that in Fig. 5.2, which is a change-over with four replicates of four treatments, but in which the treatments have been allocated completely at random to sheep-periods.

This design gives 12 d.f. for error, as opposed to the 6 d.f. attached to error in a 4 × 4 LS. However, if you intend to use four sheep and four periods, it seems foolish not to choose a LS with the additional constraint on randomization which that requires. If, when the ANOVA is completed, it is apparent that either or both of the sheep m.s. and the periods m.s. are not materially different from the error m.s., the pragmatist can add those components to the error and thus

Sheep →	i	ii	iii	iv
Periods ↓				
1	A	B	A	C
2	D	D	A	B
3	C	B	D	C
4	D	A	C	B

Fig. 5.2. A change-over design which does not allow the (easy) separation of effects due to sheep and periods because it lacks the requisite balance, but which gives more d.f. for error than a 4 × 4 Latin square.

recover the d.f. that they carry. In this way one can make a check on the assumptions that sheep variance and period variance are merely reflections of experimental error and, if the assumptions are vindicated, use the extra d.f. to improve the precision of the experiment. If either of the assumptions is not vindicated, then the error m.s. with 9 or 6 d.f. will be *smaller* than the error with 12 d.f. and the LS has turned out to be the more precise design.

The case for using a change-over design which is *not balanced* for the effects of animals or periods is clearer when the numbers of treatments, animals and replicates do not fit any convenient set of Latin squares. For example, with five treatments to compare and eight digestibility crates available, one might sensibly use two periods with eight sheep and a third period with four sheep to obtain four replications of each treatment. In this case the treatments would be allocated at random to sheep-periods.

Where there are only three treatments to be compared, three replicates of each treatment will seldom be enough since the 2 d.f. for error in a 3 × 3 LS give a large value for t (4.303 at $P = 0.05$). The usual remedy is to use two (or more) 3 × 3 squares, and there are at least three ways of arranging these. Each square can be randomized and operated independently, using $3n$ sheep where n is the number of squares, in which case the combined error d.f. will be $4n - 2$ (see ANOVA *a*, in Table 5.3). If, however, the two or more LS experiments are conducted simultaneously, it is as well to regard period effects as common to all squares and so remove only 2 d.f. for periods, rather than *2 for each square* (ANOVA *b*). A third possibility is to use the same three animals repeatedly to build up replications of the 3 × 3 design (independently randomized), in which case only the *average* effect of differences between animals would be removed (ANOVA *c*).

Balanced Latin Squares

A balanced Latin square is one in which each treatment follows every other treatment an equal number of times (the usual number is once). Figure 5.3 gives

Table 5.3. Frameworks for the analysis of variance of sets of four 3 × 3 Latin squares.

ANOVA *a.* (12 sheep in 4 independent 3 × 3 LS)	d.f.	ANOVA *b.* (12 sheep in 4 simultaneous 3 × 3 LS)	d.f.	ANOVA *c.* (3 sheep in 4 sequential 3 × 3 LS)	d.f.
Between squares	3				
Sheep (within squares)	8	Sheep	11	Sheep	2
Periods (within squares)	8	Periods	2	Periods	11
Treatments	2	Treatments	2	Treatments	2
Error	14	Error	20	Error	20
Total	35		35		35

an example of such a balanced square and if you compare it with Fig. 5.1 you will find that, whereas treatment A occurs once after B, once after C and once after D in the balanced design, it does not do this in the LS designs given in Fig. 5.1. In fact, it happens that in Fig. 5.1b treatment A always follows C and you can imagine that, if C happened to be a treatment with adverse after-effects, this would (unfairly) affect your estimate of the response to treatment A. In the balanced design (Fig. 5.3) the hypothetical bad effects of treatment C would impinge equally on treatments A, B and D.

You will find that, in a 4 × 4 square, if the design is balanced with respect to the carry-over effects of one treatment, it is balanced for all treatments. You may also discover that balance cannot be achieved in squares with *odd* numbers of treatments, except by taking *a pair* of squares into consideration.

The residual effect of treatment C for the design in Fig. 5.3 would be estimated by adding up the plots that are ringed. The residual effect of the other three treatments would be found in an analogous way. This would lead to an ANOVA as set out in Table 5.4. In this analysis, the two components labelled 'A'

Cows → / Periods ↓	Daisy	Ruby	Spot	Gert
1	B	A	D	C
2	C	B	A	(D)
3	(A)	D	C	B
4	D	C	(B)	A

Fig. 5.3. A balanced Latin square in which the treatments (A, B, C and D) each follow every other treatment once and once only.

Table 5.4. Framework for the analysis of variance in the 4 × 4 balanced Latin square illustrated in Fig. 5.3.

Source	d.f.	
Cows	3	
Periods	3	
Direct effect of treatments (unadjusted)	3	A
Residual effect of treatments (adjusted)	3	A
[Direct effect of treatments (adjusted)	3]	B
[Residual effects of treatments (unadjusted)	3]	B
Error	3	
Total	15	

would account for *the same sum of squares* as the two components labelled 'B'. The computations are performed in this way in order to arrive at estimates of the direct effects of treatments (adjusted for the residual effects of preceding treatments) and the residual effects themselves (adjusted for the direct effect of treatment). A worked example is given in Appendix 12.

Since a balanced Latin square costs no more than an unbalanced one, it seems sensible to adopt a balanced design whenever a 4 × 4 square (or a pair of 3 × 3 squares) is being planned. You must (initially) pay the price of 3 d.f. taken from the error but, if carry-over effects do arise, it is important to know about them and to isolate that component; if, on the other hand, there are no carry-over effects (that is the m.s. for adjusted residual effects is of the same magnitude as the error m.s.), the ANOVA can revert to that for an ordinary 4 × 4 LS with 6 d.f. for error. This is another case where you can have your cake *and* eat it.

Balanced Latin Squares with an Extra Period

It is possible to obtain a better estimate of carry-over effects if each treatment follows every treatment, *including itself*, an equal number of times. This can only be achieved by adding an extra period to the design, as shown in Fig. 5.4.

In principle, one could insert the *duplicated period* at any stage in the experiment but, in practice, it is safer to place it at the end, since, if something then goes wrong in period 5, the results of the first four periods can still be analysed as a balanced LS. Although the balanced Latin square with an extra period is a very sophisticated design for estimating carry-over effects, it is not commonly used in animal trials because it adds 25% to the work of what may already be a long and risky experiment and because sensible experimenters do not choose change-over designs if they suspect that carry-over effects will be present.

Although balanced designs enable some estimate to be made of residual effects of treatment, the designs discussed do this only for the immediately

Cows → Periods↓	Daisy	Ruby	Spot	Gert
1	C	D	B	A
2	A	B	C	D
3	B	A	D	C
4	D	C	A	B
5	D	C	A	B

Fig. 5.4. A balanced Latin square with an extra period, arranged so that each treatment follows every treatment once.

following period and on the assumption of an *additive* effect. That is, our analysis assumes that the carry-over effect of treatment A from period 1 to period 2 is the same as its carry-over effect from periods 2 to 3 and 3 to 4. This brings us back to the underlying assumption about parallel lactation curves. To the extent that different cows (if given the same treatment) do not show parallel curves this will confound estimates of the treatment response *and* of the carry-over effects. A switchback design, which is described below, avoids the assumption of parallel lactation curves but does not estimate or remove carry-over effects.

Balanced LS designs suffer from all the disadvantages listed as points 2, 3 and 4 on p. 45.

Switchback Designs

In a switchback experiment an animal is first given treatment A, then B and then switched back to A again. This is a very powerful way of being sure that the responses are due to the treatments applied, particularly if, at the same time, another animal is receiving a B–A–B pattern of treatments. However, they are only suitable for testing treatments where there is a prompt response to the application of the treatment and no carry-over effect once the treatment is stopped or changed. Figure 5.5 illustrates an experiment, using only two cows, which leaves little room for doubt that treatment B stimulates extra milk yield (at least in the short term).

In most cases it would be prudent to have more than one pair of cows for such an experiment, but the precision of switchback designs is such that they can give clear results with a very much smaller number of animals than a continuous trial. This is because not only does each animal act as its own control (as in other change-over designs) but also the time trend is removed independently for each animal. The flavour of the analysis is that the response to treatment in Gert is given by the quantity $[B_2 - \frac{1}{2}(A_1 + A_3)]$ and in Daisy by $[\frac{1}{2}(B_1 + B_3) - A_2]$. These

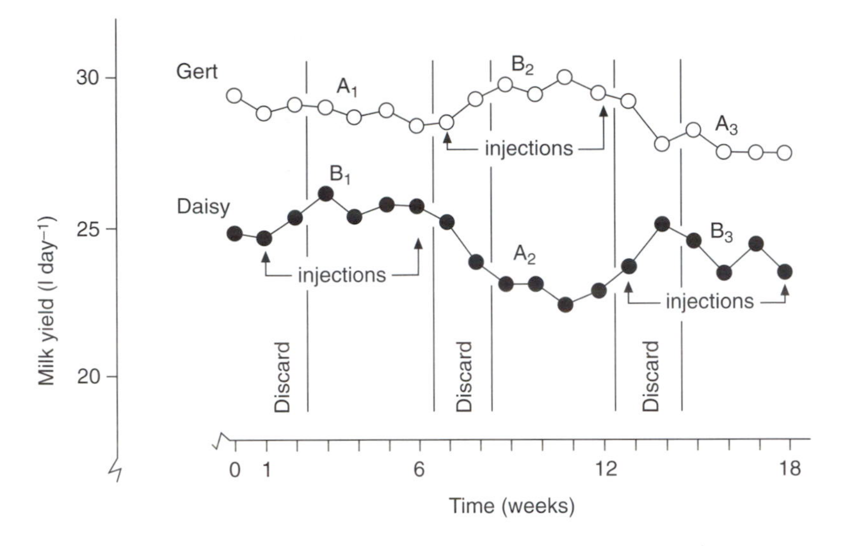

Fig. 5.5. A switchback experiment in which treatment A is the control and treatment B is a weekly injection of bovine somatotrophin.

quantities automatically remove the effect of a linear[a] change in yield with time for each cow. There is no assumption that the changes are parallel for the two cows (or even that they are in the same direction). The assumption made is that the performance expected of each cow in period 2, if the treatments had *not* been changed, can be estimated by linear interpolation between periods 1 and 3.

If you wish to use a switchback design to compare more than two treatments, this can be done by using a *pair* of animals for *each pair* of treatments. Thus, the comparison of four treatments would require a minimum of 12 animals (six pairs) to estimate A–B, A–C, A–D, B–C, B–D and C–D. The analysis of such designs has been described by Lucas (1956).

When to Use Change-over Designs

Although change-over designs can greatly improve the precision of an experiment and so reduce the number of replicates required and the total costs, there are many circumstances in animal experimentation where they are not appropriate. The main circumstances in which we should *not* use a change-over design are:

1. Whenever we wish to measure the long-term, cumulative effect of treatments on animals.

[a] If it is necessary to consider a curvilinear trend in performance during the course of an experiment, this can be allowed for by using four periods with an A–B–A–B pattern of treatment.

2. Whenever we suspect that the effect of a treatment on animal performance will persist after we have removed the treatment or changed to a new one.

As one or other (often both) of these conditions apply to the great majority of experiments involving whole animals, it is not surprising that the circumstances in which a change-over might be appropriate are rather few and far between. Table 5.5 gives some examples of traits and treatments which are, and are not, suitable for incorporation in change-over experiments.

The most valuable use of change-over trials is in the preliminary screening of treatments to see whether there is *any* response or to discover which of several treatments gives the greatest response. In this way a modest number of animals can be employed to screen a relatively large number of treatments, allowing the experimenter to follow up by designing a continuous trial capable of evaluating the long-term effects of a few treatments that look particularly promising.

Table 5.5. Some examples of traits and treatments which are and are not suitable for incorporation in change-over experiments.

Change-over may be appropriate	Change-over usually not appropriate
Effect of diet on milk composition	Effect of energy or protein intake on milk yield
Effect of diet on egg weight	Effects of energy or protein intake on rate of lay
Effect of some hormone treatments	Effects of photoperiod on reproduction
Digestibility and balance trials	Effects of any treatments on growth
	Effects of management practices on mastitis

Summary

1. Change-over designs give *greater precision* than continuous trials but cannot measure the cumulative effects of treatments over a long period.
2. Even short-term responses may be falsely estimated by a change-over trial if there is a substantial *carry-over effect* from one period to the next.
3. *Balanced Latin squares* allow the estimation of carry-over effects and are to be preferred to unbalanced Latin squares. But even a balanced LS assumes there is no animal × period interaction.
4. *Switchback designs* allow for some animal × period interaction (i.e. independent linear time trends in different animals) but assume no carry-over effects.
5. Change-over designs are useful for preliminary screening of large numbers of treatments but form a *dangerous base* from which to draw *final conclusions*.

Reference

Lucas, H.L. (1956) Switch-back trials for more than two treatments. *Journal of Dairy Science* **39**, 146–154.

Exercise 5.1

A Latin square design was used to test the effect on egg weight of including molasses in the diet of laying hens at four concentrations (0, 70, 140 and 210 g kg^{-1}). Four groups, each of 48 birds, received each diet in turn for a period of 4 weeks. Data for the first 2 weeks after changing diets were discarded. Table 5.6 shows mean egg weight for each group on each diet (based on weighing eggs in bulk on weekdays in weeks 3 and 4 in each period). The Roman numerals in brackets alongside the data give the period in which the results were obtained.

1. Carry out an analysis of variance of these data.
2. Do the figures justify a conclusion that the inclusion of molasses in the diet depresses egg weight? If so, and supposing that smaller eggs are undesirable, what level of molasses can be used in laying hen diets without causing *any* depression in egg weight?

Table 5.6. Mean egg weight (g) for groups of 48 pullets.

	Molasses in the diet (g kg^{-1})				Group totals
	0	70	140	210	
Group A	55.4 (II)	55.1 (III)	53.6 (IV)	53.5 (I)	217.6
Group B	55.0 (IV)	56.1 (I)	54.8 (II)	53.9 (III)	219.8
Group C	55.2 (III)	52.9 (IV)	53.8 (I)	54.1 (II)	216.0
Group D	53.1 (I)	54.4 (II)	53.0 (III)	51.1 (IV)	211.6
Treatment totals	218.7	218.5	215.2	212.6	865.0
Treatment means	54.68	54.63	53.80	53.15	

(The corrected total sum of squares = 22.6575)

Chapter 6

Pens and Paddocks

In the preceding chapters, the experimental units discussed have been either individual animals or (in Chapter 5) animal-periods. However, the exercise at the end of Chapter 5 presents data from a trial in which the 'unit' was a group of 48 laying hens, probably housed in a block of adjoining cages. Since four replicate groups were allocated to each diet, there were clearly 192 chickens in that experiment. It may seem that there should be lots of d.f. for error, and yet the error m.s. in the LS analysis had only 6 d.f. Why did we not use data from *individual animals* in this case to improve the precision of the experiment? This raises an issue which is frequently misunderstood (and misreported in published papers) about the analysis of data when treatments have been allocated to animals penned together in *groups*.

Groups of Animals in Pens

Individual feeding of animals in nutritional trials is sometimes possible, but the facilities required for this are specialized and expensive. In many cases, and especially with pigs and poultry, it is simpler to place a number of animals in a pen together and to give all of them a single diet. Replication is then achieved by having more than one pen on each treatment. Figure 6.1 illustrates a pig experiment laid out with pens arranged in randomized blocks.

The ANOVA for this experiment takes the standard form for an RCB design, with 6 d.f. for error variance amongst pens.

However, there might, for example, be eight pigs in each pen, making a total of 96 pigs on the experiment, so what has happened to the 95 d.f. for the variance *amongst pigs*?

The answer is that, assuming that we have recorded individual pig performance (which is likely to be true for liveweight gain but *not* for food intake), we can analyse the between-pig variation, but it does not provide an appropriate experimental error. A full analysis of individual pig records would be as shown in Table 6.1.

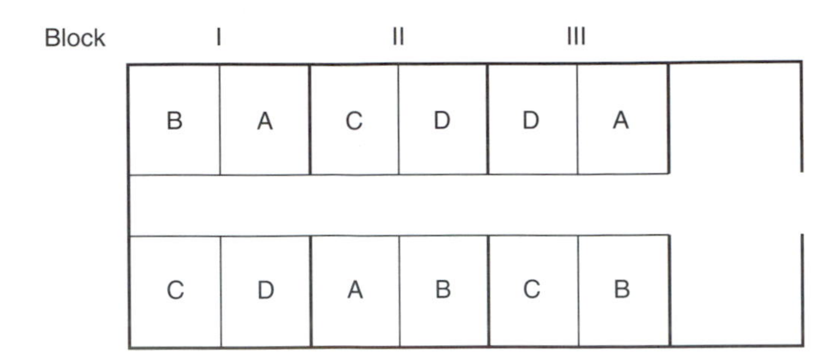

Fig. 6.1. Plan of 12 pens in a piggery in which an experiment with four treatments (A, B, C and D) has been laid down, the treatments being allocated at random to pens within three blocks.

For purposes of testing questions about treatments, the bottom half of this analysis is irrelevant and unnecessary. We only need to know the mean performance for each pen to conduct the essential ANOVA. Notice that the treatment m.s. must be tested using error *a*, the *between-pen error*, and *not* error *b*, the *within-pen error*. Why can we not use the variation amongst pigs as the basis of error in this design? The short answer is that we did not allocate treatments at random to *pigs*. We allocated treatments to *pens*. The dishing out of treatments did not involve 95 independent decisions, it involved 11 decisions (since once you have put a treatment label on 11 of the pens the label for the twelfth pen is already determined).

If you are not convinced by this statistical argument, consider what would happen if performance in one pen were to be affected by some extraneous event. Perhaps the student in charge one weekend neglects to check all the drinking bowls and a valve sticks so that the pigs in one pen get no water for 36 h. This fault is corrected on Monday morning but, in the meantime, all pigs in that pen have suffered a check to their growth. If we regard the pig as the experimental unit, the data will show that 8 out of the 24 pigs receiving that treatment had lower growth rates and this will seem a fairly well-replicated result, leading to

Table 6.1. Full ANOVA for the pig experiment illustrated in Fig. 6.1, assuming that each pen contains eight pigs.

Source	d.f.
Blocks	2
Treatments	3
Error *a* (amongst pens)	6
Total (amongst pens)	11
Error *b* (amongst pigs within pens)	84
Total (amongst pigs)	95

the false conclusion that the treatment tested in that unlucky pen was inferior. If, on the other hand, we regard pens as the experimental units, the unlucky pen will look like one aberrant result, not replicated in the other two pens receiving the same treatment, and we shall write it off as experimental error.

In fact, error *a* is made up of two components. One is the variation between pigs within pens (which is estimated directly by error *b*) and the other is the variation in pig performance which arises from them being in different pens. In a very well-designed piggery and in a very well-conducted experiment, the latter component may be zero: but we are not entitled to assume that and experience teaches us that pen variance is often substantial. Even if, in a particular case, pen effects are zero (which would mean that error m.s. *a* equals error m.s. *b*) we are still not entitled to treat the experiment as one in which pigs have been individually allocated to treatments.

Is there any point in conducting the full ANOVA, based on individual records, in a pen experiment of this type? Provided you do not fall into the trap of using error *b* instead of error *a* to test treatment effects, no harm is done; and the benefit from conducting the full analysis is that it gives you an estimate of pen variance (i.e. error a − error b). This can lead you to raise questions about your experimental facilities and procedure which may help to improve the precision of future trials. However, as we have said before, the full analysis is unnecessary for judgement of the treatment effects.

It is possible to keep animals in pens and yet to allocate treatments individually, but, for nutritional trials, this requires devices for individual feeding within the pen. The alternative to this is to place animals in separate pens, which is often cheaper and easier to manage. One should not, however, attempt to place day-old chicks in individual cages (most of them will die of starvation) and if sheep are penned individually the partitions should allow animals to see each other, otherwise they will spend all their time trying to climb out. Figure 6.2 illustrates a piggery in which individual feeders have been installed, so that pens are no longer the experimental units, but have become blocks. The ANOVA for the design in Fig. 6.2 is shown in Table 6.2.

In the case of laying hens in battery cages, it is possible to have individual cages with separate feed hoppers, which would allow individual feeding but, even when the facilities are like this, it is more usual to designate groups of chickens in adjoining cages to be fed on the same ration, because this reduces and simplifies the work and also reduces the risk of mistakes. Where the hens are assigned to treatments in groups like this, it is essential to analyse the data on the basis of group means, not individual performance, even though the primary recording of egg production may have been done on a cage by cage basis.

Keeping Records of Individuals

Is there any advantage in keeping individual records of performance where the experimental unit is a group? One can envisage three situations. First, there are

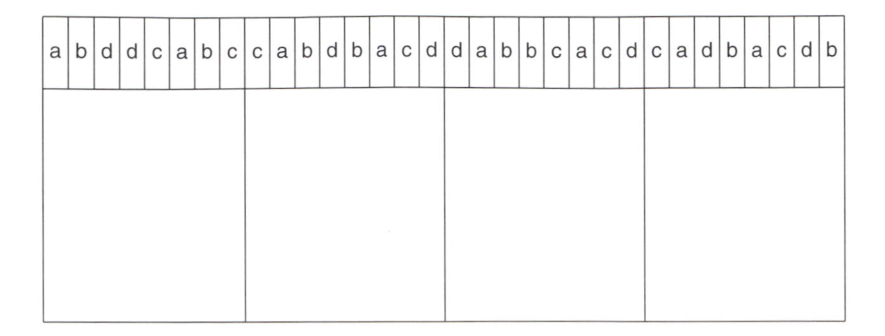

Fig. 6.2. Plan of four pig pens in which the use of individual feeders allows the comparison of four diets (a, b, c and d) within pens (two pigs allocated to each diet in each pen).

cases where the individual record is the only practical route to obtaining the mean performance of the group. This would usually be the case with pig weighings. Although it might be technically feasible to weigh the whole pen of pigs, most operators find that they can more conveniently weigh the pigs one at a time and then add the results together. Secondly, there are cases where the individual weighing, though technically feasible, does not constitute a sensible policy. An example would be the estimation of egg weight for groups of hens arranged in individual cages. You could weigh every egg separately and record it against the number of the cage from which it was collected: but there is no point in adopting this approach when it is so much easier to gather the eggs from each group onto a tared receptacle and to weigh them in bulk. Thirdly, there are cases in which, operationally, it does not make much difference whether primary recording is done by individuals or for the group. An example would be the weighing of small chicks. If these are housed in a group in a brooder compartment or in a floor pen, they first have to be caught and put in a box. You could then quite quickly weigh the chicks individually, but an obvious alternative is to weigh the box full and weigh it again when the chicks have been returned to their pen. The argument in favour of individual weighing in this case is that, if one chick in the group is sickly and subsequently dies, its records can be removed from the group mean throughout the trial, given that you have

Table 6.2. Analysis of variance framework for the experiment illustrated in Fig. 6.2.

Source	d.f.
Pens (= blocks)	3
Diets	3
Error	25
Total (amongst pigs)	31

identified individual chicks when you weighed them separately. It is also easier to make appropriate adjustments to food intake records to remove the effects of ailing or dead chicks where there are individual data available for weekly weight gains.

Grazing Trials

The question of replication in grazing experiments sometimes causes difficulty, though it should not. Figure 6.3 shows the layout of a grazing trial in which eight steers have been allocated at random to each of three paddocks. The areas of the paddocks were adjusted so as to provide three stocking rates, which are the treatments in this experiment.

This might look like a trial with eight replications and, indeed, one could conduct an ANOVA as shown in Table 6.3.

The trouble with this analysis is that the error measures only variation amongst steers *within paddocks* (7 d.f. amongst eight animals in each paddock), whereas the comparison of stocking rates is being made *between paddocks*. There is an unspoken assumption with this design that there are no territorial differences and that, had the paddocks all been made the same size, they would have supported exactly the same growth rate from the animals grazing them.

Fig. 6.3. A grazing trial using three paddocks to test the effect of three stocking rates on growth of steers at pasture (this is *not* a valid experiment).

Table 6.3. ANOVA for the trial portrayed in Fig. 6.3.

Source	d.f.
Paddocks	2
Error	21
Total	23

This assumption is both unwarranted (since experience teaches us that there are always differences in crop growth between different patches of ground, no matter how uniform the soil or the sward may look) and, what is worse, *untestable*. The whole principle of scientific investigation is that we judge the differences between experimental units treated differently in the light of observed differences in the performance of units treated alike. Here we have no basis for saying how different the results from two paddocks at *the same* stocking rate might have been.

The remedy, of course, is that we must *replicate the paddocks*. The question of how many paddocks will be needed depends on gathering some information about the variability encountered in such experiments and this information is likely to be location specific. You can imagine that pastures are more variable in regions where herbage productivity is low (because of low rainfall, high altitude or poor soils) than where sown pastures are irrigated and fertilized to give high yields. Figure 6.4 shows a possible design for a set-stocking experiment on lowland pasture in a well-watered part of the world.

You will notice that this trial could have been laid out as a 3 × 3 Latin square, but that would have given only 2 d.f. for error. The experimenter in this case decided not to sacrifice the extra 2 d.f. that would be needed to control for

1.6 ha 8 steers	3.2 ha 8 steers	2.4 ha 8 steers	Block I
2.4 ha 8 steers	3.2 ha 8 steers	1.6 ha 8 steers	Block II
2.4 ha 8 steers	1.6 ha 8 steers	3.2 ha 8 steers	Block III

Fig. 6.4. Design for a grazing trial with three stocking rates compared in three randomized complete blocks.

variation between columns, but arranged his blocks so that they ran along the contour of the land. The ANOVA for this design is given in Table 6.4.

Grazing experiments are sometimes planned to measure the effects of set stocking (that is, animals are allocated to their paddocks for a whole grazing season) but can also be designed to measure the effects of rotational grazing on animal performance and on pasture. In such trials, the animals are often allocated either for a fixed number of days, or until the herbage has been reduced to a predetermined level, and then moved either to another paddock which is due for grazing or to a holding area, until such time as the sward has regrown. The measurement of animal productivity in such trials is usually represented as a number of grazing days, as it becomes difficult to associate a particular amount of weight gain with each of the treatments being investigated.

Coefficients of Variation for Groups

The question may be asked: 'If I have an estimate of the CV for a trait measured in individual animals, what will be the expected CV for pen means based on a specified number of animals?' The answer is that the *minimum* variation for group means can be derived from the usual formula:

SEM = $s/\sqrt{n}$, where n is the sample size.

However, the variation of pen means will contain not only variation due to animals but also variation due to *pen effects* and thus the CV to be assumed for planning experiments using pens of animals will be *larger than* CV/$\sqrt{n}$ (where CV is that applying to individual measurements and n is the number of animals in each pen). The amount of pen variation which should be allowed for can only be estimated by collecting data from relevant experiments, preferably in the circumstances to be used for the trial. Moreover, pen variation, because it includes the influence of things that go wrong (and *something* goes wrong in every experiment) will show more variation from trial to trial than the variation amongst animals (within pens or blocks), which tends to be rather consistent for traits such as growth rate and yield.

Table 6.4. ANOVA for the design illustrated in Fig. 6.4.

Source	d.f.
Blocks	2
Stocking rate	2
Error *a* (amongst paddocks)	4
Amongst paddocks	8
Steers within paddocks (error *b*)	63
Total (amongst steers)	71

What we can assert as a generalization is that an experiment involving, say, 12 groups of 8 animals (11 d.f. in total) will have much better precision than a trial with 12 individuals (11 d.f. in total), but nothing like the precision of a trial with 96 individuals (95 d.f. in total).

This means that the planning of group-based experiments involves more uncertainty than trials where the individual is the unit, but this disadvantage is normally offset in practice by the use of many more animals in total than could be treated as individuals. As the number of animals in each experimental group gets larger, the contribution to between-pen error which comes from $s/\sqrt{n}$ (where s is the individual variation) gets smaller and smaller and the error is more and more made up of pen effects. This means that the calculation of an expected CV from individual data is completely unreliable when you come to experiments that have dozens or hundreds of animals in each group, as may be the case in trials with chickens. In these cases, data are needed from previous experiments conducted in the particular trial facilities.

Summary

1. In any experiment where treatments have been allocated to groups of animals in pens or paddocks, the pen or paddock is the *experimental unit*, not the animal.
2. *Individual feeding* arrangements within pens make it possible to allocate nutritional treatments independently to individuals within the pen.
3. A grazing experiment in which the number of paddocks equals the number of treatments is an *unreplicated* experiment.
4. Estimation of an appropriate CV for experiments with groups of animals is tricky.

Exercise 6.1

In an experiment designed to investigate the vitamin A requirement of growing chicks, 100 day-old male broiler chicks were allocated at random to ten pens of ten chicks. Two pens were allocated at random to each of five diets. Liveweights at 4 weeks of age are recorded in Table 6.5.

1. Compute an appropriate SEM for each treatment.
2. Set out a full ANOVA showing variation amongst chicks within pens, as well as variation amongst pen means.
3. Compute a linear regression of chick weight on dose of vitamin A. Does the ANOVA indicate significant deviations from a straight-line regression?
4. Was it necessary to weigh each chick separately in this experiment?

Table 6.5. Weights of chicks (g) at 28 days in an experiment comparing five doses of dietary vitamin A.

	Added vitamin A (i.u. kg^{-1})									
	0		500		1000		2000		4000	
Pen no.	6	9	3	10	4	7	1	8	2	5
	211	237	352	397	420	375	467	444	450	437
	284	274	393	372	344	340	366	507	427	452
	300	214	284	365	407	358	421	480	480	355
	238	208	329	360	442	397	484	529	413	416
	283	240	398	384	395	375	444	467	429	489
	242	276	343	427	397	437	445	396	506	434
	290	283	376	440	324	406	404	458	467	441
	283	255	367	354	418	302	380	363	444	391
	221	233	315	313	462	313	432	454	392	423
	*	193	378	358	*	421	433	473	435	468
Pen totals	2352	2413	3535	3770	3609	3724	4276	4571	4443	4306
Pen means	261.3	241.3	353.5	377.0	401.0	372.4	427.6	457.1	444.3	430.6
Treatment means	251.3		365.2		386.7		442.3		437.4	
s.s.	8848.0	8476.1	11854.5	12362.0	15318.0	18024.4	12014.4	21304.9	9964.1	12982.4
Variances (s^2)	1106.00	941.79	1317.17	1373.56	1914.75	2002.71	1334.93	2367.21	1107.12	1442.49
Standard deviation	33.3	30.7	36.3	37.1	43.8	44.8	36.5	48.7	33.3	38.0

* One chick died.

Chapter 7

Factorial Designs

It is not unusual to plan experiments that investigate two or more factors simultaneously. A simple example would be a trial evaluating the improvement in digestibility due to alkali treatment applied to straw samples chosen for their range of indigestible fibre contents. We already known that alkalis such as caustic soda can improve the digestibility of fibrous forages fed to ruminants. The hypothesis to be tested is that the degree of improvement in digestibility depends in some way on the initial quality of the untreated straw. We might, for example, select eight straw samples, either on the basis of their chemical compositions or by using evidence from a previous *in vivo* or *in vitro* digestibility trial. We might then choose three levels of alkali treatment (e.g. 0, 20 and 40 g NaOH kg^{-1} dry matter).[a] The usual way of proceeding would then be to test all 8×3 combinations of straws and alkali levels, making 24 treatments to be evaluated. This is a rather large number of treatments to test by means of a full digestibility trial in sheep or cattle, but is a manageable number for *in vitro* evaluation in the laboratory or an *in vivo* trial using dacron bags incubated in the rumens of fistulated animals.

In this experiment we would say that we are investigating two factors (straws and alkali), the first factor at eight levels (note that 'levels' does not always imply a numerical description: it can also imply qualitative differences or multidimensional differences, e.g. 'breeds' which could differ in a number of quantifiable ways) and the second factor at three levels.

Suppose that such a trial has been carried out with two replicates of each treatment, the replicates representing successive batches, so that batches are complete blocks in this case. The ANOVA for this experiment would then be constructed as in Table 7.1.

The component labelled 'Straws' is often referred to as the 'main effect' of straw, which amounts to asking how much the straws differ when we consider their digestibility *averaged across* all three levels of the alkali treatment.

[a] Notice that zero is regarded as a level and that in a trial where the treatments were simply 'none' and 'some' we should refer to that as two levels of a factor. However, when you come to enter data for computer analysis, it makes no difference whether you label three levels of treatment as 0, 1 and 2 or 1, 2 and 3.

Table 7.1. Outline of the preliminary analysis of variance of an 8 × 3 factorial design in two blocks.

Source	d.f.	
Blocks (= batches)	1	
Treatments	23	
Straws		7
Alkali treatment		2
Interaction		14
Error	23	
Total	47	

Table 7.1 is labelled the 'preliminary analysis' because it is unlikely to answer all the questions the experimenter wishes to ask. The next steps will depend on whether or not there is interaction between the factors being investigated.

Factorial Analyses with No Interactions

If, in the example above, there is no interaction, this means that the effect of alkali treatment on straw digestibility is the same in all eight straws. The interpretation of the experimental results is then simple. The experimenter may wish to ask further questions about the main effects (e.g. 'Is there a linear regression of digestibility on dose of alkali?' or 'Are all the straws different from each other or is it useful to divide them into groups in some way?'), but he can perfectly well talk about the effects of alkali on *all* the straws tested and equally well talk about the differences between the straws without having to specify whether they were alkali-treated or not.

What do we mean by 'no interaction'? Should this statement read 'no *significant* interaction'? Well, no – it should not. This is another case in which you should not feel wedded to the 5% probability level and especially so when there are many degrees of freedom for the interaction term. If, in the example in Table 7.1, we find that the probability of the F-ratio for interaction is around the 10% level, we should be very wary of the conclusion that the experiment can be interpreted entirely in terms of its main effects. More probing is required to see why the interaction m.s. is greater than the error, and this may lead to sensible hypotheses about what has happened, which can then be made the subject of proper statistical testing (see below). It could be, for example, that the straws differ significantly in their response to the first dose of alkali (20 versus 0 g NaOH kg^{-1}) but show smaller and not significantly different responses when the 40 g dose is compared with the 20 g dose.

Factorial Analyses When Interaction Is Present

If there is a significant interaction, or if the interaction term is *suspiciously large*, the analysis in terms of main effects and interactions should be discarded. If the response to alkali treatment depends on which straw we apply the treatment to, there is not much point in talking about the *average* response to NaOH treatment. Equally, if differences in digestibility of the straws depend on whether or not they have been treated, there is little value in discussing the straws in terms of their average digestibility across a number of treatments (including no treatment).

Whenever interaction is demonstrated or suspected, the treatments s.s. should be subdivided in some other way to yield meaningful components. The form of that subdivision will be dictated by the data in each case and so there are no general rules of procedure, other than:

1. Look carefully at the treatment means and draw some graphs (or perhaps Venn diagrams, if you know about those) to represent your results pictorially.
2. Seek for an interpretation of the data in ordinary language which seems to make sense to you.
3. Set down your interpretation as a set of formal hypotheses.
4. Challenge those hypotheses with appropriate statistical tests.

Two for the Price of One

Factorial designs are most often used when the experimenter is concerned directly with the question of interaction. In the example above, the researcher would not have been interested in designing a single-factor experiment. He already knew that caustic soda was capable of improving straw digestibility. He also knew that his straw samples (untreated) had a wide range of digestibility values. The question being investigated was whether the response to alkali treatment was in some way dependent on the initial digestibility of the untreated straws. This question can only be answered with a factorial combination of treatments and the experiment is essentially one investigating the interaction.

It is, however, possible to adopt a factorial design where there is no expectation that the factors being investigated will interact and the experimenter(s) rather hope that they do not. The objective in these cases is to carry out *two* experiments *for the price of one*.

Suppose that one research worker wishes to compare the value of four different forages (grass silage, maize silage and mixtures of the two in the proportions 1:2 and 2:1) fed to dairy cows throughout the period of winter housing. He calculates that he will need a minimum of 15 cows per treatment and puts in his request to be allocated 60 cows for the coming season. This is an expensive experiment in terms of the capital facilities and labour required to house the cows and record their individual feed intakes. Suppose also that, at the same research station, another worker wishes to test the efficacy of a newly available orally active form of bovine somatotrophin (BST). She decides that she needs three treatments: a control (no treatment), cows which receive weekly injections (a treatment for which there is already much data showing its efficacy) and the oral treatment which is to be evaluated. She puts in a request for 36 cows to conduct this experiment.

Now 96 cows for these two experiments may be beyond the available resources of this station (taking into account other requests for animals already granted to senior research staff), so the Director proposes that the two factors should be combined in a full factorial design and applied to 60 cows. This would make $4 \times 3 = 12$ treatments and there will only be five cows on each treatment. Both contenders argue that this is not sufficient replication. However, the Director points out, quite rightly, that the young man will have the 15 cows on each forage treatment which he asked for and the young lady will have 20 cows on each of her three treatments, which is more than she had asked for. Both research workers then complain that there may be interactions between forages and hormone treatments and that this will spoil their results. The answer to this is in two parts.

First, interaction between these two factors does not seem inherently likely and the trial may turn out to have an interaction term no larger than error. In that case, each researcher has a good chance of a successful experiment and they can report their results independently, even in separate papers in different journals if they wish.

Secondly, there may, contrary to expectation, be some interaction between forages and hormone treatments. Supposing that to be the case, would the young lady like to specify which of the four forage treatments she would wish to use in her experiment on BST? She will probably reply that she does not care or cannot choose which diet to use as a background for her experiment, except to say 'whatever is normal practice'. As normal practice in the country where she works includes feeding cows on grass silage and maize silage and on mixtures of the two, this specification does not resolve the argument. Equally, the nutritionist may be challenged to say whether he wishes his trial to be conducted with or without BST treatment of the cows. If this happens to be a country where half the dairy cows are regularly injected with BST (or a country where BST is not yet licensed for use, but might be within the next decade) he may have some difficulty in answering this question. The more you think about it, the more it seems reasonable to ask the question about orally active BST using cows which are receiving *a range* of practical diets and to ask the question about forages of cows which *are* and *are not* being dosed with BST. If there *is* some interaction between these factors it would be better to know about it, as it would be rather important in making recommendations about forage diets and about BST.

Now it is true that, if interaction is expected and is the *object* of investigation in this trial, five cows per treatment is an inadequate number. What number would be adequate depends on the nature of the interaction hypothesized.

However, nobody anticipates an interaction; if, in truth, there is one, then the trial is likely to give some indication of that. What is certain is that both main effects are likely to be adequately tested by an experiment with 60 cows and at least two research papers should emerge. This is yet another case where you can have your cake and eat it.

The use of factorial designs to answer two *unrelated* questions using the same set of animals is a much underexploited trick in animal experimentation.

Factorial Designs with Unequal Replication (Split Plots)

In most animal experiments where two or more factors are investigated, all the factorial treatment combinations will receive equal replication. There are a few circumstances, however, where this is either not possible or not convenient, and you are then likely to end up with a split-plot design.

Figure 7.1 shows a field experiment laid out in split plots, where one factor (ploughing versus direct drilling) has been tested on large plots of ground, which have then been split into subplots for comparing the second factor (fertilizers).

In field experiments there are *two* reasons why a split-plot design might be preferred to a design in which factors are equally replicated on plots of uniform size. Either the design is *chosen* to give greater precision to the factor which is allocated to the subplots (e.g. if the main plot factor in Fig. 7.1 were cereal varieties, the design might have been chosen to give maximum precision to the fertilizer comparison (and interactions) on the grounds that the comparative yields of the varieties were already well known): or the split-plot design is *dictated* by physical considerations. In Fig. 7.1 the design is a split plot because ploughing and direct drilling can only be done over plots large enough to accommodate the machines, whereas fertilizers can be applied accurately by hand over much smaller areas.

In animal trials, split-plot designs usually arise because *circumstances dictate* it. If you have read Chapter 6, you will realize that much of what follows

Block	I		II		III		IV	
	P	D	P	D	D	P	D	P
	a	c	c	b	d	c	a	d
	c	b	a	d	b	a	b	c
	d	a	b	c	a	b	c	a
	b	d	d	a	c	d	d	b

Fig. 7.1. Plan of a field experiment testing the effects of ploughing (P) versus direct drilling (D) in combination with four levels of nitrogen application (a, b, c and d) in a split-plot design.

is going over the same ground, but it bears repetition. An example of a split-plot design dictated by circumstances is given in Fig. 7.2, where an experiment with laying hens has been planned to investigate the interaction between environmental temperature and dietary nutrient concentration. The question is whether putting more energy (and protein and minerals in due proportion) into the diet helps to overcome the adverse effects of heat stress on laying performance. Four diets have been formulated and these are to be tested at three temperatures (comfortable, mild stress and severe stress). The temperature comparisons can only be made by allocating a separate controlled-environment room to each temperature and, with three temperatures to test, we shall need a minimum of six rooms (note that if this experiment is conducted in three rooms it is an unreplicated trial and therefore not a scientific experiment, just like the grazing trial illustrated in Fig. 6.3). If the dietary comparisons were to be replicated on the same basis as the temperatures, we would need a minimum of 24 rooms but, fortunately, this extravagance is unnecessary. The obvious solution is to compare the diets *within* the rooms, which turns it into a split-plot design.

The ANOVA for the design in Fig. 7.2 is given in Table 7.2. Notice that there are two distinct errors, one for testing treatments applied to whole rooms (temperatures in this case) and the other for testing effects compared within rooms (diets).

If this design is analysed as though it were a factorial with equal replication (24 plots with 23 d.f., 11 d.f. for treatments and 12 d.f. for error) the error would be a compound of between-room and within-room variance. This error is *too small* for legitimate testing of the main effects of temperature and *too large* for the efficient testing of dietary main effects and the crucial interaction.

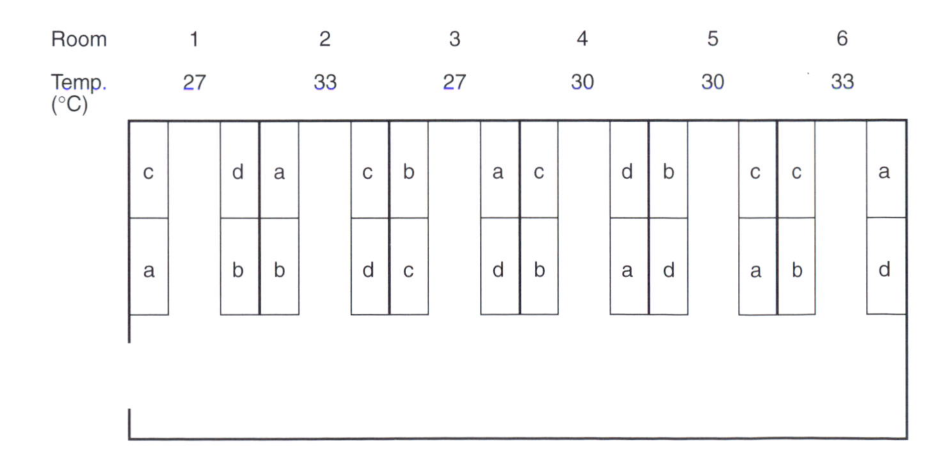

Fig. 7.2. Plan of an experiment testing three temperatures (allocated to six rooms at random) in combination with four diets (a, b, c and d) allocated at random within rooms in a split-plot design.

Table 7.2. Outline of the analysis of the design illustrated in Fig. 7.2.

Source	d.f.
Temperatures	2
Error (a) between rooms	3
Rooms	5
Diets	3
Temperature × diets	6
Error (b) within rooms	9
Total	23

The need for replicated rooms (or replication in time) when investigating the effects of environmental factors on animal performance is worth reiterating, since this prerequisite of good design is often found to be lacking in published reports. Students (and, sadly, their supervisors) often argue that the rooms to be used are *identical* and therefore differences in performance observed must be due to the environmental factors deliberately imposed. Sadly, this argument does not wash. If you build rooms to identical specifications, stock them with animals and run them with *the same* control settings in all rooms, there will be differences in animal performance between the rooms due to a whole variety of causes, including drift from the intended environmental specifications, accidental effects (e.g. a malfunctioning drinker) and non-specific disease. Experience shows that the between-room variance is almost always larger than the within-room variance and it is therefore essential that rooms are replicated so that an appropriate error can be calculated for testing treatments applied to whole rooms. There may or may not be additional factors investigated within the chambers but, if there are, this calls for a split-plot analysis which separates the within-room and between-room variances.

Summary

1. Factorial combinations of treatments can give valuable information about the *interaction* of two or more factors.
2. When interaction is found to be present, the discussion of '*main effects*' becomes *irrelevant* and the standard factorial analysis should be abandoned.
3. *Two* experiments can be done *for the price of one* by combining unrelated questions in a factorial design.
4. Some factorial designs involve unequal replication and these will usually require a *split-plot* analysis with two (or more) error terms.
5. Where one of the factors to be investigated is the response to (controlled) environments, this will require *replicated rooms* and a split-plot analysis.

Exercise 7.1

The data in Table 7.3 are taken from an experiment in which chicks were fed a basal diet of cereals and noog meal, with the object of finding the first limiting amino acid in diets using noog meal. Supplements of lysine, methionine and cystine were tested in all possible combinations. Each diet was fed to six groups of 20 chicks from 7 to 21 days of age.

Table 7.3. Liveweight gain (g) for chicks fed on eight different starter diets ($n = 6$ groups of 20 chicks).

Diet	Treatment totals	Treatment means
Control	920.3	153.4
+ Lys	909.7	151.6
+ Met	970.2	161.7
+ Cys	979.0	163.2
+ Lys + Met	1000.8	166.8
+ Lys + Cys	992.0	165.3
+ Met + Cys	960.1	160.0
+ Lys + Met + Cys	994.2	165.7
Grand total	7726.3	

A preliminary ANOVA (Table 7.4) shows that there are significant differences amongst the treatments.

Table 7.4. ANOVA for weight gain in chicks (7–21 days old) (a partial summary of the data is given in Table 7.3).

Source	d.f.	s.s.	m.s.	F
Blocks	5	225.3427	45.0685	2.21
Treatments	7	1360.2581	194.3226	9.52^{***}
Error	35	714.4797	20.4137	
Total (pens)	47	2300.0805		

1. Subdivide the treatment s.s. into main effects and interactions and test whether these are significant.
2. Carry out any further tests that you think might be useful in helping to answer the question: 'What is the first limiting amino acid in noog meal?'
3. Summarize the results of the experiment in a short report, without using technical expressions such as 'Main effects'.

Chapter 8

Assumptions Underlying the Analysis of Variance

When we conduct a routine analysis of variance we are making three assumptions. These are that:

1. the data being analysed represent a normally distributed continuous variable;
2. the errors about the various treatment means are homogeneous; and
3. an additive model is appropriate.

We will discuss number 3 (additivity) briefly at the end of this chapter and devote Chapter 13 to a discussion of data that are not normally distributed. The main topic to be dealt with immediately is the assumption about *homogeneous treatment variances*.

You may like to note that, if you use a computer package to analyse your data, it may have built-in routines to check on all three of these assumptions and to issue warnings if your data fail the tests. However, not all software packages are as cautious as this and you should be aware of the potential difficulties yourself, particularly if you use a pocket calculator or a very 'simple' computer program for ANOVA.

Homogeneity of Variances

The business of doing an ANOVA and arriving at a *pooled error m.s.* assumes that the variation of replicate values about the treatment mean is the same for all treatments (except for small variations due to sampling). This is often true in animal experiments, and especially in those trials where we measure productivity with traits such as growth rate, milk yield or egg yield. It is unusual in such trials for the differences between treatment means to be more than 5–15% and, on that scale, altering the level of yield does not usually alter variation amongst replicates very much.

However, when we measure a trait that can change its mean level by fourfold or tenfold, it is unlikely that the variability of replicate responses will remain the same.

Consider the data in Fig. 8.1. Not surprisingly, plasma luteinizing hormone (LH) concentration has been increased dramatically by injections of gonadotrophin releasing hormone (GnRH), but also the SEM has increased roughly in proportion to the mean concentration of LH. In this case the variance was calculated separately for each treatment and *not* derived from an ANOVA. The pattern of response is clear enough and testing of differences between individual treatments, although hardly necessary, could be done by *t*-tests if desired (refer to Appendix 2 for the SED between two means which do *not* have equal variances).

However, we may have a more complex experiment in which there are different groups of subjects, perhaps in different phases of the reproductive cycle, or different agents to be applied and compared (e.g sheep GnRH versus a synthetic decapeptide). Now the question to be asked might be whether the slopes of the responses are similar and then the differences in SEM for the several treatment means become a nuisance.

The Logarithmic Transformation

What is happening in the results illustrated in Fig. 8.1, and in many other similar cases, is that the SEM is increasing roughly *in proportion* to the mean.

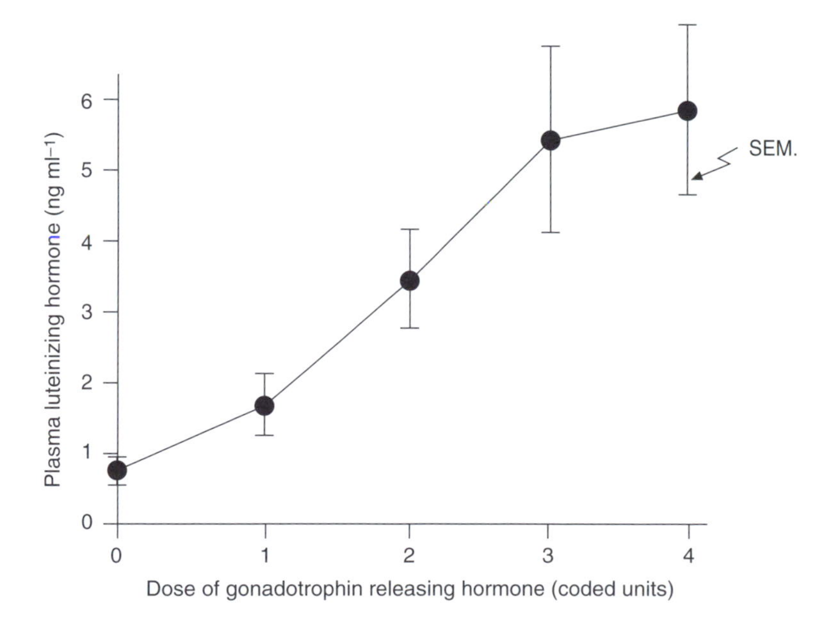

Fig. 8.1. Response to injection of gonadotrophin releasing hormone (means ± SEM).

In other words, the CV remains constant. When this is the case the remedy is simple. By transforming plasma LH values to *logarithms* of the plasma concentration, we obtain homogeneous treatment variances and so can carry out an ANOVA and give a pooled value for the SEM. We can also conduct tests within the ANOVA (such as comparison of regression slopes obtained with different agents) without fear of drawing spurious conclusions. Note that if a pooled variance were to be used for the data in Fig. 8.1, without first making the logarithmic transformation, this would lead to an SED which would be unduly large for separating treatment responses at the bottom of the response curve but too small for valid testing of effects at the top end. This can lead to serious mistakes when comparing two curves which apparently diverge at high dose rates.

It doesn't matter whether you use *natural logarithms* (to the base e = 2.7183) or logarithms to the base 10 as a means of transforming proportional data to an additive scale. What matters is that you define which scale you used when you report the results, so that the reader can convert the data back to a more familiar scale if he or she wishes. Note, however, that the SEM or SED which you quote cannot simply be transformed back to the original scale of measurement and still have meaning. All statistical judgements must be made using the transformed (logarithmic) data if that was the basis of your ANOVA.

Whenever you find that treatment responses are very large (say >100% of control values) you should *begin* with an assumption that the data will need transformation before analysis. This assumption needs to be checked, but visual examination of the records before and after transformation is often sufficient. Computer packages have more sophisticated methods.

Testing for Homogeneity of Variance

If you find it necessary, you can always check whether the treatment variances are homogeneous (that is, can they reasonably be regarded as samples from a single population of variances?). If there are only two treatments, calculate the variances separately for each treatment (after removing effects due to blocks or other features of the design, if these are present) and compare the treatment variances by dividing the larger m.s. by the smaller one. The result is an F-ratio which you can compare with the tabulated values in Appendix 25 to discover the probability that variances as different as these two might have occurred by chance.

If there are more than two treatments, again calculate the variances separately for each treatment and run a quick check by comparing the largest and the smallest of these variances as an F-ratio. If this leaves room for doubt as to whether the whole set of variances could be regarded as belonging to a single normal distribution, you should then apply Bartlett's test, which is described in Appendix 15.

Further Examples

Before leaving this discussion of the scale effect (i.e. bigger means having bigger SEMs in direct proportion), it may be useful to point out that, although the problem most frequently arises in dose–response experiments examining physiological traits, there are other examples. For instance, if you conduct a lactation trial with two breeds that differ in their milk fat content as widely as Holsteins and Jerseys, it is not wise to analyse data for milk fat without first checking that the variances are homogeneous across breeds. Another case was reported by Morris and Njuru (1990) who tested responses to five protein levels fed to broiler chicks and to cockerels of a layer strain. The broilers had a mean growth rate three times that of the layer-strain chicks and, not surprisingly, also had a much larger SEM for some traits (but not all: food intake was actually more variable in the layer-strain chicks). The remedy adopted in that case was to conduct all ANOVAs separately for the two breeds. This did not provide a formal examination of the breed × diet interaction, but as it was obvious that the response pattern of the two breeds differed in many respects, that was no great loss.

Other Transformations

The logarithmic transformation is the appropriate one if you are simply dealing with a 'scale effect' – the SEM being proportional to the mean. However, other distortions of variance do occur (although much less frequently) and may require other transformations.

A *square root* transformation may be appropriate if you find yourself dealing with *areas*, such as leaf area (commonly estimated by taking length × breadth as an index) or the cross-sectional area of histological preparations such as muscle fibres or seminiferous tubules. Here the argument is that if a linear measurement, such as length or diameter, is responding in a manner that does give homogeneous variances across treatments, then the areas measured (or estimated) will vary in proportion to the square of those linear measurements, since the area of a circle = πr^2 and the area of a constant shape of length l is kl^2. Taking square roots of the data will resolve these problems. However, it is not a *rule* that the variability of diameters or lengths will remain constant across all treatments and so the data should always be examined to see what kind of variation is present before rushing into a particular transformation.

Counts of things such as ticks on cattle are prone to show heterogeneous variances, although here the heterogeneity may be between breeds showing different susceptibilities, or pastures with different exposure rates, rather than between treatments applied. Such counts often give better conformity to the assumptions needed for an analysis of variance after the data have been transformed. Sometimes the square root transformation is appropriate in these cases and sometimes a logarithmic transformation works better. With computing power it is an easy matter to investigate the alternatives.

Percentages are usually approximately normally distributed when the data lie between 20% and 80%; but with values close to zero or close to 100% the data are apt to be badly skewed. The remedy for this is to use the *angular transformation*:

$$p = \sin^2 \phi$$

where p is the percentage you need to transform, expressed as a proportion (i.e. on a scale from 0 to 1), and ϕ is the corresponding angle (in degrees) the square of whose sine is p. This transformation can be made by using a prepared table in a good statistics text or a computer package. You will find that percentages of 2 and 4 transform to 8.13° and 11.54° whereas 52% and 54% become 46.15° and 47.29°. The tails of the percentage distribution are stretched out and central values are compressed.

Examples of data that commonly need angular transformation are *mortality rates* (for *groups* of animals: individual mortality is an all-or-none trait: see Chapter 13) and *fertility* of eggs, measured either for groups at a single hatching or in individual hens over an extended period.

Additivity

An assumption underlying routine analysis of variance is that the several elements in our experimental design affect the responses to be measured in an additive way. For a simple RCB design the underlying model is:

plot yield = general mean + block effect + treatment effect + random error.

This supposes that if, for example, block 2 gives higher yields than block 1, this effectively *adds on* a given quantity to all the yields in block 2. It does not assume that all the yields in block 2 will be x% better than the corresponding treatments in block 1. That would be a *multiplicative* effect and requires a different analysis. Logarithmic transformation of the data may sort this problem out, but note that we are then making an assumption that random errors are greater in blocks (and treatments) with higher mean values, and this is not always true. Where there is a (foreseen) possibility that treatment responses will be of different magnitudes in different blocks, it is a good idea, as pointed out in Chapter 2, to increase the number of replicates allocated to each treatment in each block from one to two (or more). This will allow you to compute the variance amongst treatments which are in *the same* block and use this to test the reality of the block × treatment interaction (see Tables 2.9 and 2.10).

Another example where non-additive effects may easily occur is in factorial designs; but here the additivity is routinely tested by computing the interaction term. For example, the interaction term in Table 7.1 tests the extent to which the effect of alkali treatments is not the same in all eight varieties of straw.

Where an interaction between two factors is detected and you suspect that the response is due to *proportionality*, a logarithmic transformation of the data may be helpful. For example, Table 8.1 gives some data for food intake as affected by temperature in two breeds of hen. Analysis of the data showed a highly significant interaction between breed and temperature, but the experimenter suspected that this was just a scale effect. The Rhode Island Reds ate more food and showed a greater numerical reduction in food intake due to temperature, but in both breeds the drop in intake between 15° and 30°C was about 25%.

When the data were transformed to a logarithmic scale the significant interaction disappeared, confirming that proportional responses to temperature were the same in the two breeds. In this case, there was no significant heterogeneity of variance due to breed differences on either the linear or the logarithmic scale. The transformation was not needed to validate the assumption about homogeneous errors which underlies the analysis of variance, but was useful as an aid to *interpretation* of the results.

Table 8.1. Mean feed intake from 30 to 60 weeks of age for hens of two breeds maintained at four constant environmental temperatures.

	Feed intake (g per bird day^{-1})		Log_{10} feed intake	
Temperature (°C)	White Leghorn	Rhode Island Red	White Leghorn	Rhode Island Red
15	114	128	2.057	2.107
20	108	120	2.033	2.079
25	101	111	2.004	2.045
30	86	95	1.934	1.978
SEM	0.71	0.71	0.0025	0.0025
Significance of interaction	$P < 0.01$		$P > 0.20$	

Summary

1. It is important to check, before doing an ANOVA, that your data are *normally distributed*, that the treatment *variances* are *homogeneous* and that block and treatment effects are *additive*.
2. These assumptions seldom fail in trials measuring animal productivity, because treatment responses are modest, but will frequently fail where physiological responses are measured, because these can involve *large changes in mean level* for different treatments.
3. The most common case encountered is that in which the SEM for different treatments is *proportional* to the mean. These cases call for transformation of the data to a logarithmic scale.
4. A square root transformation is often helpful where the trait measured is an *area* or a *count* of numbers.

5. Failure of the *additivity* assumption may occur when there are large differences between *blocks* or in *factorial* experiments. These cases are sometimes resolved by a logarithmic transformation of the data.

Reference

Morris, T.R. and Njuru, D.M. (1990) Protein requirement of fast- and slow-growing chickens. *British Poultry Science* **31**, 803–809.

Exercise 8.1

The data in Table 8.2 come from an experiment in which sheep were given *ad libitum* access to diets based on hay or grass silage or maize silage. Plasma lactic acid concentrations were measured in samples of venous blood. It seems that lactic acid concentrations were both higher and more variable in the sheep receiving silage. Are the differences in variance such that we should not undertake a routine ANOVA which pools them together?

You will need to calculate the variance separately for each treatment. However, as some of the variation down the columns in Table 8.2 can be attributed to block effects, the first step is to remove this portion of variance. The last column of Table 8.2 lists the mean influence of each block on the plot yields, obtained by subtracting the general mean (0.5667) from each block mean. These block effects must be *subtracted* from the plot values in corresponding rows to obtain Table 8.3.

You can now calculate the variance within each treatment and then test whether those variances are homogeneous.

Table 8.2. Plasma lactic acid concentrations (mmol $litre^{-1}$) measured in 18 sheep receiving three diets. Animals were assigned to six blocks on the basis of their voluntary feed intake recorded in a preceding trial.

	Diet					
Block	Hay	Grass silage	Maize silage	Block totals	Block means	Block effects
1	0.21	0.54	0.82	1.57	0.5233	−0.0433
2	0.36	0.27	0.97	1.60	0.5333	−0.0333
3	0.19	0.36	1.29	1.84	0.6133	+0.0466
4	0.29	0.44	0.86	1.59	0.5300	−0.0367
5	0.24	0.22	1.48	1.94	0.6467	+0.0800
6	0.18	0.33	1.15	1.66	0.5533	−0.0133
Totals	1.47	2.16	6.57	10.20		0.0000
Means	0.245	0.360	1.095	0.5667		

Table 8.3. Plasma lactic acid concentrations adjusted to remove block effects.

	Diet				
Replicate	Hay	Grass silage	Maize silage	Totals	
1	0.2533	0.5833	0.8633	1.6999	check that
2	0.3933	0.3033	1.0033	1.6999	row totals
3	0.1433	0.3133	1.2433	1.6999	are now all
4	0.3267	0.4767	0.8967	1.7001	the same
5	0.1600	0.1400	1.4000	1.7000	apart from
6	0.1933	0.3433	1.1633	1.6999	rounding

Chapter 9

Dose–Response Trials

In Chapter 3 we met the argument that, for experiments exploring dose–response relationships, the statistical examination of differences between treatment means is not only a waste of time, but can be very misleading. Consider the data in Fig. 9.1.

You will see that the yield of milk protein responded as might be expected to intakes of protein which resulted from four dietary treatments. In such circumstances the fact that differences between the second, third and fourth treatments are not significant (at $P = 0.05$) is irrelevant. There are no reasonable grounds for doubting that milk protein output is responding in a continuous curvilinear manner to the increasing supply of protein. The best estimate of response is given by a fitted curve, *not* the treatment means. There is plenty of room for argument about the most appropriate model for the curve (see the discussion below), but no grounds for arguing that the response ceases at a protein intake of 1250 g day^{-1}.

Given a set of data from a dose–response experiment or from a time series (e.g. growth as a function of time), it is often apparent that some curvilinear analysis is called for, but how do we choose an appropriate curve? The most important advice here is to think about the biology of the response and to avoid relying too heavily on statistical testing of alternative models. With most data sets from animal experiments it is not possible to discriminate between alternative models on the grounds that one is a significantly better fit than rival candidates that might be considered. Looking again at Fig. 9.1, a straight line accounts for most of the variance between the treatments and a curve, although a better fit in the sense that it leaves a *smaller* residual variance, does not actually result in a *significant* ($P < 0.05$) reduction in error variance. To use this result as an argument for a linear response is just as absurd as saying that there is no response beyond a protein intake of 1250 g day^{-1}.

It is often difficult to say from first principles, what form of curve we should expect, but there are some occasions where underlying theory gives rise to a reasonable hypothesis about the shape of the response expected. Some common examples are given below.

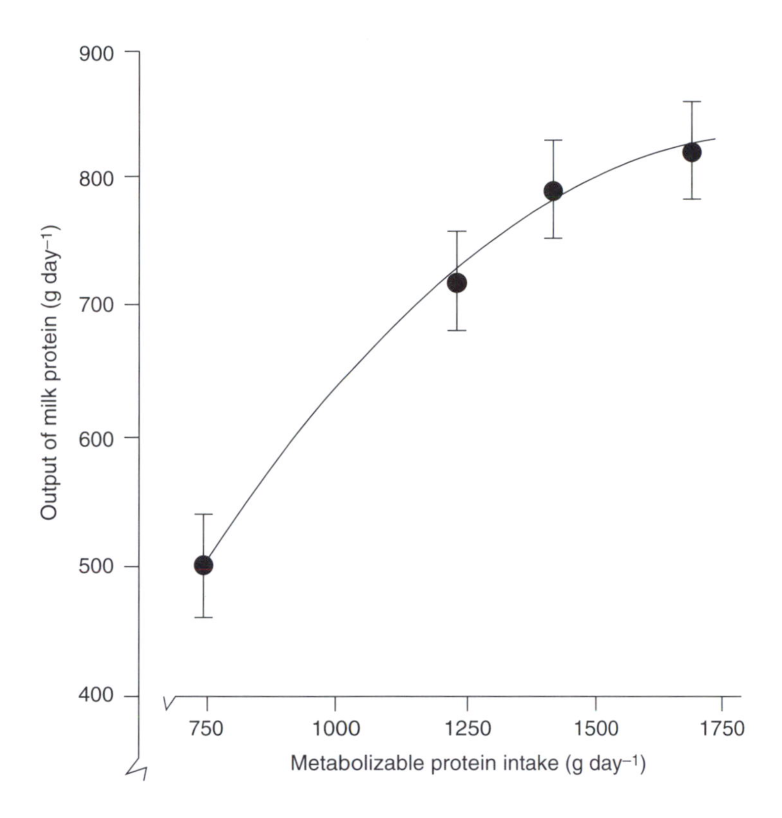

Fig. 9.1. Data from an experiment in which dairy cows were offered four diets varying in protein content. The difference between the first and second treatments is significant ($P < 0.05$) but differences between the second, third and fourth treatments are not. The fitted curve is of the form: $y = a + bx + cx^2$. Vertical bars indicate ± SEM.

Shapes of Response Curves

Exponential responses, as illustrated in Fig. 9.2, occur commonly in biology but are rarely seen in the results of animal experiments. A population growing at a constant rate (e.g. adding 1% each day to the existing population) shows *exponential growth*, but this concept only applies to the growth of individual animals in the early embryonic stages. Animal growth from conception to maturity (under non-limiting conditions) is well represented by a *Gompertz* function, which describes a double exponential curve of the form:

$$W_t = A{\cdot}\exp[-\exp\{-B(t - t^*)\}]$$

where exp means 'e to the power of'; e = 2.718, the base of natural logarithms;

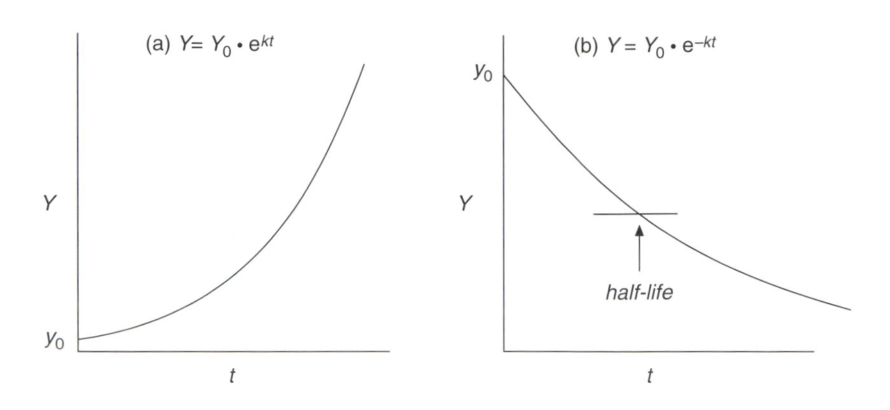

Fig. 9.2. Exponential curves of (a) growth and (b) decay. These curves can be fitted by regressing the logarithm of the response (y) on time (t), since log $y = a + kt$.

W_t = weight at time t;
A = the asymptote (i.e. weight at maturity);
B = the rate of decline (day^{-1}) in relative growth rate;
t^* = age at maximum growth rate (i.e. the point of inflection on the growth curve).

A full discussion of the Gompertz equation is beyond the scope of this book but a useful account of its application to modelling growth will be found in Emmans (1989). The Gompertz equation is particularly valuable for modelling feed intake and growth, including growth of body components, and is often the best model to use if you want to fit smooth curves to treatment data for purposes of interpolation. This would allow you, for example, to calculate feed conversion ratios and body composition at a series of intermediate slaughter *weights* for treatments that had involved slaughter at fixed *times*. Computer packages are now widely available for fitting Gompertz functions to experimental data.

Exponential decay (Fig. 9.2b) applies to radioactivity but is not typical of losses in animal populations. These are more often approximated by the ogive of a normal distribution. Figure 9.3b shows an ogive of the type often obtained when you plot gestation interval, age at puberty or age at death as a cumulative frequency. Such traits are usually normally distributed (barring a few awkward outliers) and the response is defined easily by calculating the mean and standard deviation of the sample, assuming that you have an interval from conception to birth or an age at puberty or death for each individual. The cumulative curve then represents the normal probability integral as given in standard statistical tables.

Sometimes, however, you may only know, for a series of fixed times of observation, the *proportion* of animals that have or have not undergone a specified event such as ovulation. If the exact time of ovulation cannot be observed in individual animals, examination of sets of animals by laparoscopy or serial slaughter at selected times will give the proportion of animals that have

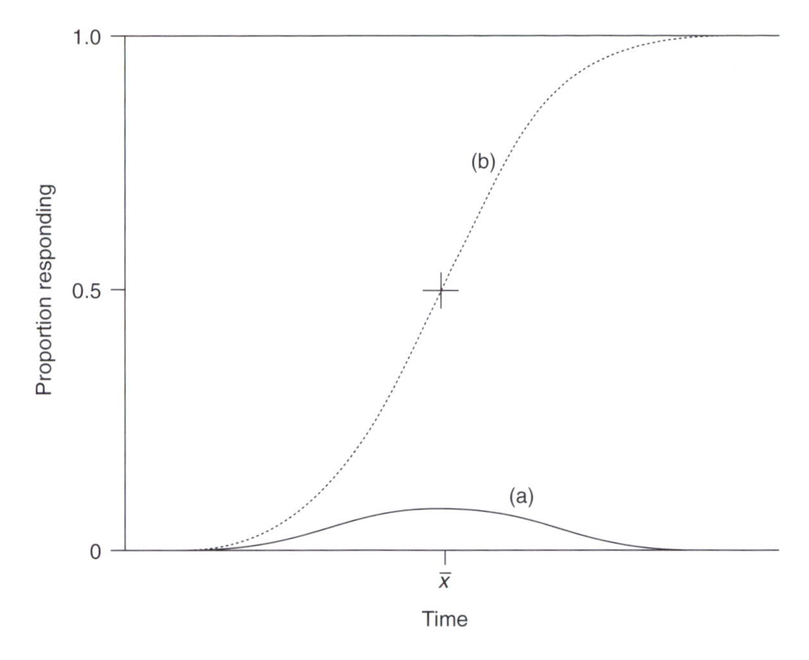

Fig. 9.3. Curves representing the normal distribution: curve (a) gives the proportion responding during any given day (or week); curve (b) (called the ogive) gives the total proportion that has responded by the end of any given day (or week).

already ovulated at each time. These data can be subjected to *probit analysis* to provide an estimate of the mean time of ovulation and its standard deviation in the group as a whole, and will allow comparison of means between different treatments. If you think you need to do a probit analysis, consult a statistician or a textbook on statistics.

None of the curves we have so far described in this chapter will fully represent the typical dose–response relationships encountered in most animal experiments if the input being studied is a nutrient, a toxin or an environmental factor (including space). All these input variables are characterized by responses up to some limit (or from some limit in the case of toxins), beyond which no further response is obtained. These responses are *asymptotic* and we need to look carefully at the problem of handling asymptotic relationships.

Asymptotic Responses

Most problems investigated in animal experiments have underlying relationships of the form illustrated in Fig. 9.4. Output is expected to improve with increasing inputs in a manner which may initially be linear, but becomes curvilinear as the ceiling is approached. Eventually, there is a plateau, where the input being

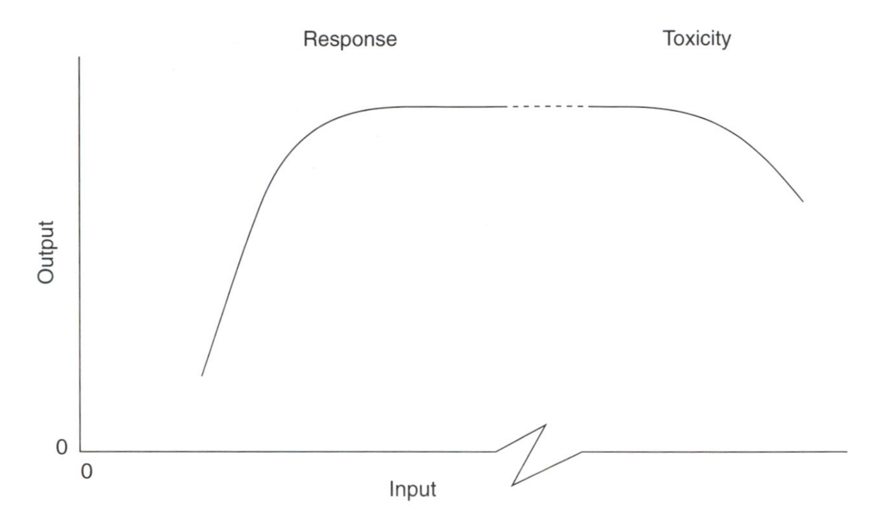

Fig. 9.4. A general model of response to increasing inputs, applicable to many animal experiments. Note that the model implies that some positive input is required before any output can be obtained, although, in practice, it may be difficult or unethical to investigate the responses to very small inputs. Some inputs will lead to 'toxicity' at high doses, but others do not.

investigated is no longer limiting performance and there is therefore no response to further increases in input. At the high end of the scale, there may or may not be a reduction in output when the input is supplied in great excess. Some vitamins and minerals, for example, are toxic at high doses, but the dietary concentration that causes acute toxicity is generally tens or hundreds of times greater than the amount needed for normal metabolism. Some inputs, such as space per animal, do not show any decline in output even though we continue to increase the supply up to any limit we care to think of. Other inputs have natural upper limits which cannot be exceeded, such as the limit imposed by appetite on energy intake.

There are some cases where the plateau is short, or even reduced to a point, giving a curve which rises to a maximum and immediately starts to fall again. The addition of a free amino acid to a pig or poultry diet will often give responses in growth and feed efficiency of this type. Although the positive and negative responses are probably not strictly symmetrical, these cases can often be adequately represented by a quadratic equation. However, experiments are not commonly designed to investigate both positive and negative (toxic) effects of a factor at the same time. Thus the experimenter limits his choice of inputs and, as a result, is almost always seeking to fit a curve which goes smoothly up to (or down from) a presumed plateau.

An essential point to understand about these asymptotic relationships is that asking at what input value the response has ceased (or begins, in the case of toxicity) is an unhelpful question. It is in the nature of an asymptotic response that the point where it ceases to climb (very slowly) and first becomes level is extremely hard to define. The answer depends upon the mathematical model chosen and, for most models, is technically 'when the input reaches infinity'! A much better question to ask is 'What equation best represents the relationship between these two variables?' If you are wise, you will think of this question not only in relation to data from your most recent experiment, but in relation to the whole body of data available to you. In most cases this will lead you to search for a curvilinear asymptotic model, which will, in turn, enable you to make statements about optimum or recommended doses. *Optimum doses* can be defined where the inputs and outputs are convertible to money units; *recommended values* must be used if the output is difficult to price (e.g. the weight of a chick at 2 weeks or the apparent freedom from disease of an animal population).

Fitting Straight Lines as a Compromise

It is possible to fit *two straight lines* to a set of dose–response data and this has the apparent advantage that it provides an exact point estimate of 'the optimum'. The problem with this approach is that, if the response is truly curvilinear, the method consistently underestimates the dose at which output is maximized and bears no settled relation to the dose at which profit is optimized. Moreover, the concept of a two-straight-line response is misguided for *anything other than an individual animal.*

It may be true that an individual animal responds to some input in a linear fashion, up to some limit, beyond which there is no response. This would be represented by the pair of equations:

$$\text{for } x \leq x^*,\ y = a + bx \qquad (9.1)$$
$$\text{for } x \geq x^*,\ y = y_{max} = a + bx^* \qquad (9.2)$$

For nutritional experiments, Equation 9.1 can be seen as the usual statement that an animal has a maintenance requirement ($x = -a/b$ for $y = 0$) and a production requirement (y_{max}/b) directly proportional to its potential yield.

Two straight lines, one sloping and one horizontal, can be fitted easily to experimental data as in Fig. 9.5. If you do not have a computer package for this, estimate where you think the hinge point will fall; then divide your data into a lower set (to be fitted by linear regression) and an upper set to be fitted by a horizontal line representing the mean of those points. Calculate the residual sum of squares about the pair of fitted lines. Now repeat the exercise with the hinge point moved to the right of the nearest data point on the right and again with the hinge point moved to the left of the nearest data point on the left. One of these calculations should yield a minimum residual sum of squares (the first one, if you guessed correctly), indicating the best fitting pair of lines. The true hinge point is where these lines intersect. We shall call this a *bent-stick* model.

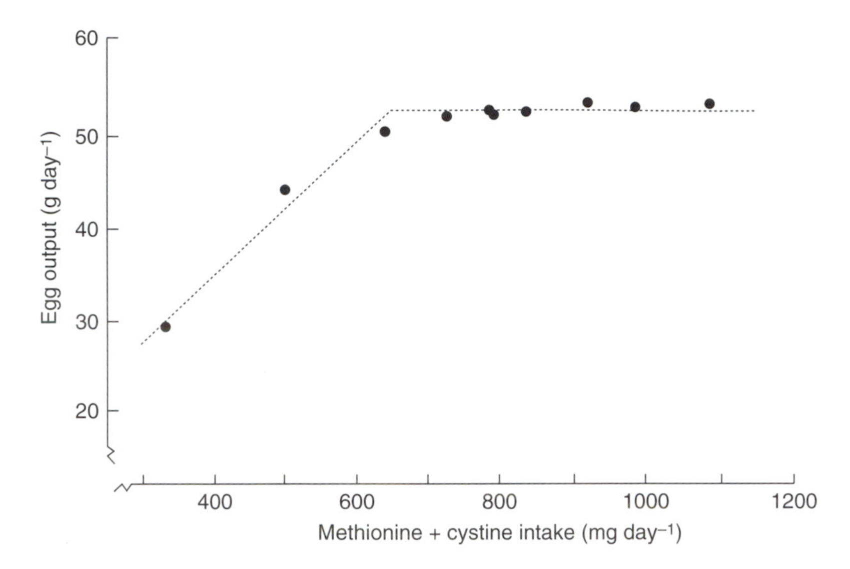

Fig. 9.5. An illustration of the bent-stick model fitted to experimental data from Morris and Blackburn (1982), in which laying hens were fed from 30 to 40 weeks of age on diets of varying protein content, methionine being the limiting amino acid in the protein mixture used.

The bent-stick model has the attraction of simplicity and (apparent) certainty about where the 'requirement' lies, but it is necessarily wrong for real situations in which we gather data from groups of animals and seek to apply the results to populations of animals. The reason for this is illustrated in Fig. 9.6.

Animals that all have the same nutrient requirement per unit of body weight and per unit of output will, nevertheless, have different response lines because they vary in size and productive potential. For responses to nutrient inputs, therefore, the bent-stick model cannot be defended for interpreting data from animal trials which necessarily use groups of variable animals, or for predicting the responses of the variable animal populations to which the results of experiments must be applied.

For non-nutrient inputs, such as space per animal or atmospheric ammonia concentration, it seems intuitively unlikely that there is a sharp end-point (or beginning, in the case of ammonia) to the response, as implied by the bent-stick model. Diminishing returns to successive increments in input seems a more realistic hypothesis for space, whether we are considering chickens in cages, pigs in pens or grazing animals at pasture.

Lest you should think that all responses have to be curvilinear, we will recall the data from Exercise 5.1. When hens were fed diets with increasing proportions of molasses, egg weight declined in a linear fashion, as shown in Fig. 9.7. Also shown in Fig. 9.7 are bent-stick and quadratic models fitted to the same data. The residual mean squares for these three models do not differ

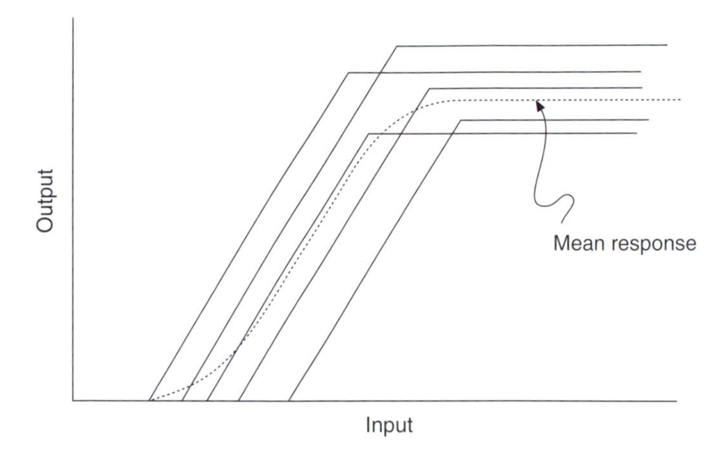

Fig. 9.6. The reason why a bent-stick model is always wrong in theory for a nutritional experiment with a group of animals. The five individuals illustrated do not vary in their efficiency of nutrient utilization (represented by the slopes of the lines), but do vary in their maintenance requirements (the intercepts) and production potentials (the plateaux).

significantly, although the smallest residual is found with the linear model.[a] In these circumstances we would be wise to choose the model that makes most sense. It is clear that incorporating molasses at 210 g kg^{-1} diet depresses egg size. Is this a toxic effect? No, it seems more likely to be a dilution effect. Because molasses adds water (as well as sugar), the energy concentration of the diet is being lowered by molasses inclusion. If this is the mechanism at work, there would be good reason to expect a continuous effect throughout the range of molasses inclusions. Therefore we adopt the linear model and conclude that there is no 'safe dose' of molasses (though small inclusions will have only small effects on egg size and these may be of no practical consequence).

Analogous arguments apply to ionizing radiation. It may be impossible to show that small doses have significant effects upon the health of exposed populations, but the evidence from larger doses indicates a linear relationship between dose and incidence of cancer. We therefore have good grounds for saying that there is no safe dose. All exposure carries some risk; but the risk from low doses is very small and it may well be worth submitting to an X-ray examination if that can help to diagnose a much more imminent threat to our health.

[a] If this surprises you, remember that fitting a curve or a bent stick takes one more d.f. than fitting a straight line. Although the more complex models necessarily have smaller residual s.s. than the straight line, they have larger residual m.s. in this case, being based on only 1 d.f. instead of the 2 d.f. for residual variance about the linear model.

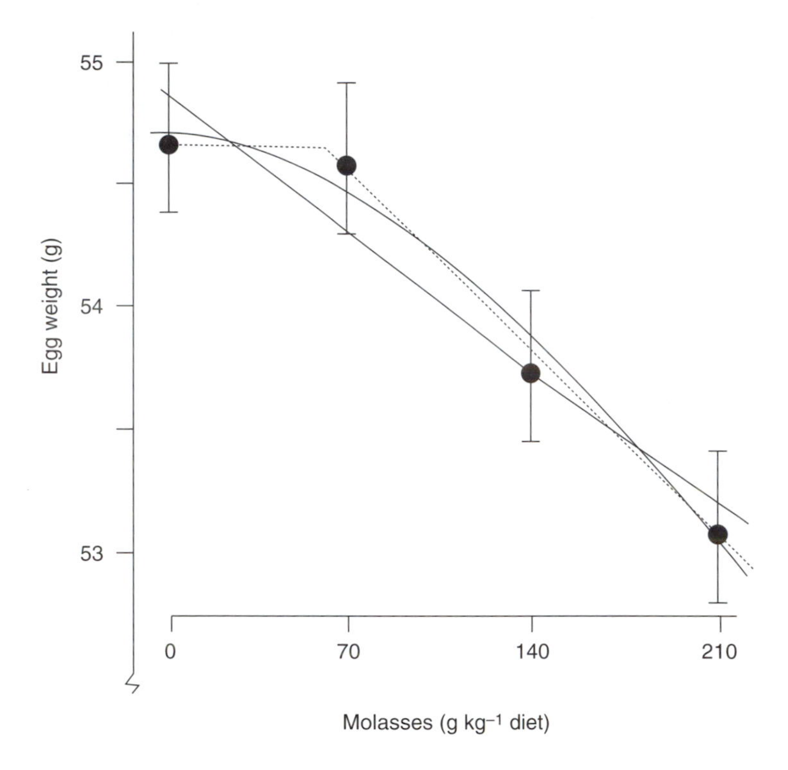

Fig. 9.7. Data from an experiment in which four levels of molasses were incorporated in the diet of laying hens (Exercise 5.1). The fitted models represent linear, quadratic and bent-stick responses. The model with the smallest residual m.s. is the straight line but, since all three models fit very well, the one that seems most likely to accord with nutritional theory should be adopted.

Simple Curves

Although the quadratic equation defines a parabola, not an asymptotic relationship, it does suffice to fit many sets of experimental results where there are a limited number of data points. With only three treatments, a parabola fits the three treatment means exactly. With four treatments, as in Fig. 9.1, a parabola is usually a good-enough fit and it is unlikely to be worth looking further unless you have some theory about the nature of the response to guide you.

So long as you are only concerned with interpolation between the experimental observations, you cannot go far wrong with a parabola fitted to four or five data points. Alternative models will give values that differ only trivially within the range of the data. Extrapolation is another matter. If you are concerned with estimating the input that gives maximum output, alternative models can give very different predictions and the parabolic model is sensitive to the range of inputs that you have chosen. This is illustrated in Fig. 9.8.

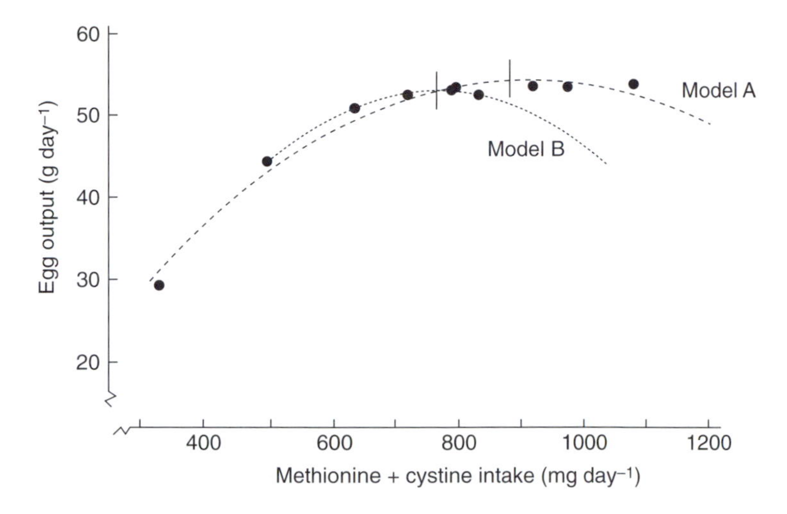

Fig. 9.8. Quadratic equations fitted to the data first described in Fig. 9.5. Model A uses all ten treatment means. Model B shows what would have happened if the trial had tested only the six central treatments. Estimated maximum outputs occur at inputs of 1000 mg day^{-1} for model A and 776 mg day^{-1} for model B. The vertical bars mark estimated optimum intakes.

If the experimenters had used six diets in the middle of the range instead of ten (still a large experiment) the estimate of methionine + cystine required for maximum output would have been altered by a factor of 0.78. Calculation of an optimum input (see Morris, 1989) yields estimates that are closer together for the two models, but still differ by a factor of 0.86.

Clearly, if the true relationship is asymptotic, a parabolic curve gives false predictions of the response to high inputs. Limiting the range of doses to avoid this problem can have a marked effect on estimates of optimum intake or maximum output. To get around these difficulties, there are at least four options. You can adapt the bent-stick idea by fitting a curve leading up to a plateau instead of a straight line and a plateau; you can use an exponential or an inverse polynomial equation; or you can adopt the Reading response model. Philosophically, the curve-and-plateau model is not very attractive, since it does not correspond with any clear biological hypothesis and, perhaps for that reason, has not been much used. The other options are described below.

Exponential and Inverse Polynomial Models

Of various exponential models that might be considered, one of the simplest is:

$$y = A - b.C^{-x}.$$

This gives a maximum value (A) to the output variable (y) when $x = \infty$.

The inverse polynomials constitute a family of equations, of which the simplest is:

$$1/y = a + bx.$$

This also reaches a maximum value of y when $x = \infty$.

Figure 9.9 illustrates these two models fitted to the Morris and Blackburn data (from Fig. 9.5).

You will notice that neither model represents the ascending part of the curve particularly well, which gives rise to a suspicion that the slope is not well estimated in the region where decisions about cost effectiveness have to be taken. Other input–output relationships (e.g. crop responses to fertilizer) may be well fitted by an inverse polynomial equation, but nutrient responses in animals often seem linear within a certain region, then pass smoothly through a region of diminishing returns to reach a plateau at a finite (and plausible) value of the input. Such a response cannot be modelled with a single equation, but needs pairs or sets of equations to cope with the regions where input is or is not limiting the response. The Reading model, which is derived from Equations 9.1 and 9.2, does just this.

The Reading Model

The most satisfactory model for the data presented in Fig. 9.5 is known as the *Reading model*. This takes as a starting point an assumption that Equations 9.1

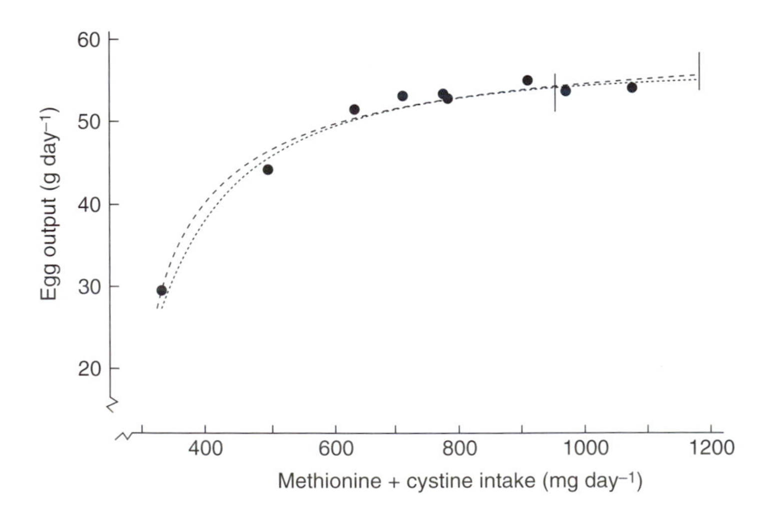

Fig. 9.9. Exponential (····) and inverse polynomial (----) models fitted to the data first described in Fig. 9.5. Estimated optimum inputs of methionine + cystine are 964 and 1184 mg day^{-1} for the exponential and inverse polynomial curves respectively.

and 9.2 are valid for *individual* animals. It then applies the argument of Fig. 9.6, saying that individuals are variable in their maintenance requirements (because they vary in size) and in their productive potentials. By assuming that the quantities W (body mass; assumed to be directly proportional to maintenance requirement) and y_{max} (potential yield) are normally distributed, and assuming that all individuals have the same net efficiency of nutrient utilization for maintenance and production, a model is developed that predicts a sigmoid curve of the type illustrated in Fig. 9.6. This curve reaches a genuine plateau when the requirements of the most demanding animal are satisfied. Although normal distributions theoretically extend to infinity in both directions, the proportion of the population lying outside ± 3 standard deviations is so tiny that it can be safely ignored. The Reading model is illustrated in Fig. 9.10.

This model was first introduced by Fisher *et al.* (1973) and the statistical theory that underlies it has been published by Curnow (1973). Examples of its application to data from laying hen trials can be found in Morris and Wethli (1978), Morris and Blackburn (1982) and Huyghebaert *et al.* (1991). The model seems to be generally useful for fitting experimental results where there is expected be a direct causal dependence of the output on the input, as would be the case for the supply of a limiting amino acid to a non-ruminant animal. It may be applicable to problems such as energy intake in the dairy cow, but here the

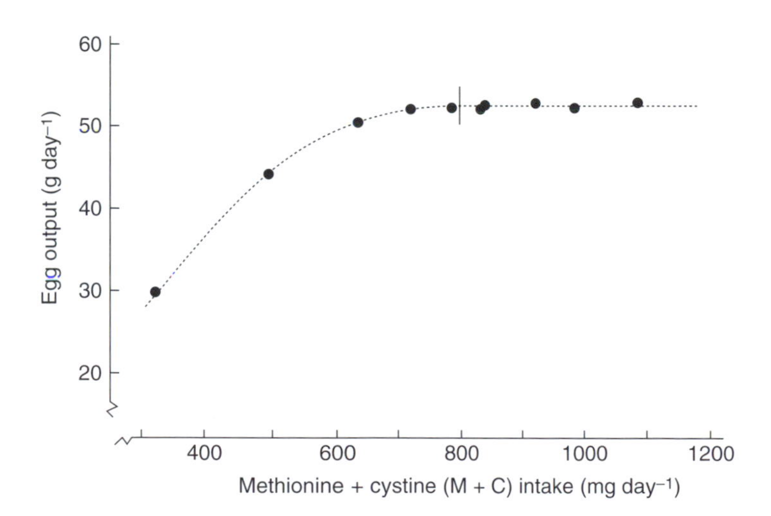

Fig. 9.10. The Reading model fitted to the data from Fig. 9.5. The underlying assumption is that individual hens have a requirement for M + C defined by M + C = $aE_{max} + bW$, where E_{max} is potential egg output and W is body mass. The population response curve is then derived from the observed normal variation in E_{max} and W. The estimated optimum input with this model (794 mg day^{-1}) should be compared with those obtained, using the same prices, from the exponential and inverse polynomial models in Fig. 9.9.

output would need to include not only milk energy but also changes in body stores. It is important that variability of yield within the experimental population is approximately normally distributed about the mean. This assumption does not hold, for example, in an ageing population of hens, where the distribution can actually be bimodal, with some individuals laying very few eggs while others are still producing six or seven eggs a week.

Two of the principal advantages of the Reading model over alternatives, which might give curves fitting the data equally well, are:

1. The Reading model will make sensible predictions of maximum output and optimum input even when the number of treatments is small or their position is less than optimal for fitting other curves. This is because the curvature of the Reading model depends only on the standard deviations of W and y_{max} and the correlation between them. The *shape* of the curve is not affected by the position of treatment means on the page.
2. Coefficients derived from fitting the Reading model correspond to real biological numbers. One of these is the input required to maintain a unit of body mass and the other is the marginal input required per unit of output. These quantities can be estimated and thus verified by other routes, including balance trials with non-producing animals to estimate maintenance requirement and determination of composition of the output to give a minimum value for the input required per unit of output.

A more extended comparison of the results of fitting different curvilinear models to the same set of experimental data can be found in Morris (1989). In order to fit a Reading model to your own experimental results, you will need the relevant computer package.

Choice of Treatments

It would seem logical to have talked about choice of treatments *before* discussing how to analyse results but, sadly, we learn by our mistakes. Now that you have read the preceding parts of this chapter you will be in a better position to appreciate how easy it is to choose an inappropriate range of treatments that will lead to no clear idea about the shape of the response curve.

The classical mistake is to say 'I know from previous work that the optimum input is in the region from about x_1 to x_2 and therefore I will concentrate my investigation in that region'. You can now appreciate that, if you have a fair idea of the shape and position of a response curve, the best way of tying it down is to explore the *whole range* of the response, so far as you are able to do so.

Figure 9.11 illustrates two designs, each intended to investigate the same input–output relationship and each using four treatments (many trials manage only three!). In Fig. 9.11a the experimenter has chosen doses that lie close to the optimum, but in 9.11b she has spaced the treatments widely. In the first case,

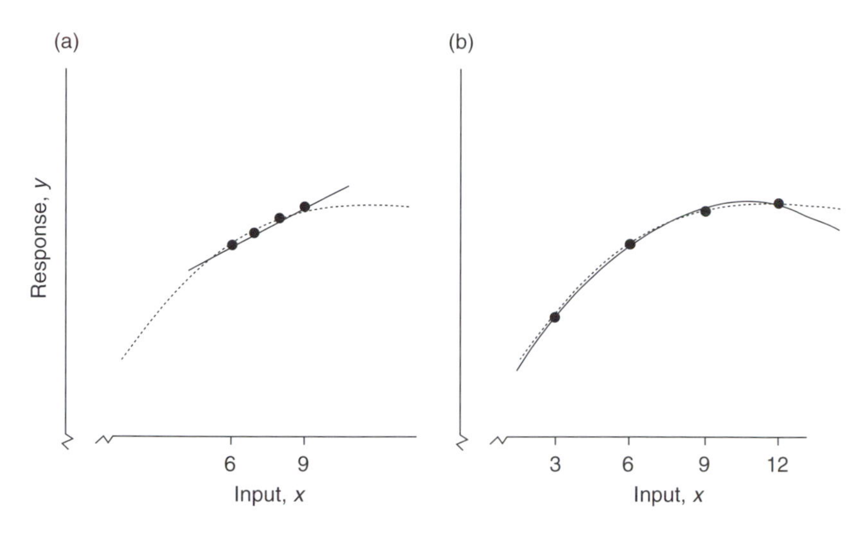

Fig. 9.11. Two designs for investigating an input–output response. In (a) the experimenter has chosen doses that focus on the area of greatest interest: in (b) she has chosen doses more likely to reveal the overall relationship between *x* and *y*. The presumed underlying relationship is indicated by the broken line. Figure (a) shows the fitted linear response, which does not differ significantly from a curvilinear one. The solid line in (b) represents a fitted quadratic equation.

there is no useful information about the shape of the response curve. It happens that a straight line fits the data rather well. If a quadratic curve is fitted it predicts maximum output at a yield that is unattainable with the genotype employed. The doses selected in the experiment illustrated in Fig. 9.11b, on the other hand, give a sensible estimate of both maximum yield and the optimum input, even if a quadratic curve is used to interpret the results. Space your treatments widely in order to obtain the best estimate of what is happening in a narrow range of commercial interest. The argument that an extreme treatment is known to be *uneconomic* is no argument at all.

Treatments involving inputs well below the optimum (and also those far above) give us vital information about the shape and position of the response curve, but these may not be pleasant treatments for the animals exposed to them. However, it is not necessary to give all treatments equal replication and this principle can be used to minimize animal distress. A smaller number of animals will suffice to fix the general position of the upward slope, whereas greater replication is needed at and near the asymptote, particularly if you wish to discriminate between alternative models of response. In the Morris and Blackburn experiment illustrated in Figs 9.5 and 9.10 only three groups (each of 48 hens) were used for the two lowest doses whereas six replicates were allocated to each of the remaining treatments.

It is thus a good general principle to *study extremes* whenever you are seeking general relationships, because these are likely to test your theories and equations most severely. If you wish to produce a general equation representing the fasting heat output of homeotherms, begin with a mouse and an elephant (and perhaps two species in between). If your equation fits that lot, it will not be far out for homeotherms in general.

Response Surfaces

This chapter has discussed only the relationship between one input variable and one output variable. Many important problems involve the simultaneous response of a system to two or more inputs. The influence of calcium, phosphorus and vitamin D on bone strength is one obvious example and the effects of dry-bulb temperature, relative humidity and insolation on feed intake and milk output would be another. These problems are usually tackled by fitting equations that define response surfaces rather than lines on a graph in one dimension only. This is a very important topic in data analysis and has even more important implications for experimental design but it is, unfortunately, beyond the scope of this book. If you find yourself confronting these problems, you should consult a statistician before beginning your experiment. Of course, that is good advice for any researcher at any time, but it is particularly important if you are to achieve an efficient design for exploring a response surface.

Summary

1. Dose–response relationships must be analysed by searching for a suitable *equation*, not by comparison of treatment means.
2. Many responses encountered in animal experiments are *asymptotic*. The response over a limited range may appear to be linear or quadratic, but a better understanding is obtained if a much *wider range* of inputs is considered.
3. Curves represented by a single equation usually do not fit asymptotic relationships ideally, although *quadratic*, *exponential* or *inverse polynomial* models may suffice for interpolation.
4. The *Reading model* is a good tool for interpreting responses where there is a direct causal dependence of the output variable on the input variable.
5. *Response surfaces* must be fitted where the output depends on two or more inputs that interact.
6. In *choosing treatments* to explore dose–response relationships, it is usually wise to select *widely spaced* treatments, including low and high extremes.

References

Curnow, R.N. (1973) A smooth population response curve based on an abrupt threshold and plateau model for individuals. *Biometrics* **29**, 1–10.

Emmans, G.C. (1989) The growth of turkeys. In: Nixey, C. and Grey, T.C. (eds) *Recent Advances in Turkey Science*. Butterworths, London, pp. 135–166.

Fisher, C., Morris, T.R. and Jennings, R.C. (1973) A model for the description and prediction of the response of laying hens to amino acid intake. *British Poultry Science* **14,** 469–484.

Huyghebaert, G., de Groote, G., Butler, E.A. and Morris, T.R. (1991) Optimum isoleucine requirement of laying hens and the effect of age. *British Poultry Science* **32,** 471–481.

Morris, T.R. (1989) The interpretation of response data from animal feeding trials. In: Cole, D.J.A. and Haresign, W. (eds) *Recent Developments in Poultry Nutrition*. Butterworths, London, pp. 1–11.

Morris, T.R. and Blackburn, H.A. (1982) The shape of the response curve relating protein intake to egg output for flocks of laying hens. *British Poultry Science* **23,** 405–424.

Morris, T.R. and Wethli, E. (1978) The tryptophan requirements of young laying pullets. *British Poultry Science* **19,** 455–466.

Exercise 9.1

The data in Table 9.1 are taken from an experiment in which 64 multiparous dairy cows were allocated to eight treatments in four randomized complete blocks. The data have been adjusted by covariance for differences in milk fat measured in days 8–14 of lactation. Four concentrates were formulated to supply 13 MJ ME and 180 g CP per kg DM, with the proportion of energy supplied from starch altered by substituting wheatfeed, sugar-beet pulp and citrus pulp for barley, wheat and cassava meal. From 21 days after calving all cows were fed 13.5 kg day^{-1} of concentrates together with hay, either *ad libitum* or at the fixed rate of 4.5 kg day^{-1}.

1. From the data given, reconstruct the ANOVA table as far as you are able. Test the significance of the main effects of concentrate type and of level of hay supplied and the interaction.
2. Find a suitable computer package and fit successively the following models relating x (= starch in the concentrate) to y (= fat content of the milk):

Table 9.1. Mean fat content of milk in weeks 4–20 of lactation of cows fed one of four concentrate diets with hay supplied in fixed amounts or *ad libitum*.

Concentrate		Milk fat (g kg^{-1})	
Number	Starch content (g kg^{-1})	Hay fixed	Hay *ad lib.*
1	429	25.6	30.3
2	319	32.0	34.8
3	231	37.1	42.0
4	141	38.5	40.8
		SEM = 2.16, 52 d.f.	

(a) a single linear regression;
(b) a pair of parallel linear regressions, one for hay restricted and one for hay *ad lib.*;
(c) a single parabolic curve;
(d) a pair of parallel parabolic curves.

Which of these models do you think best describes the data?

3. Suggest a different x variable (for which you do not have the data, but which you could calculate if you had access to the full experimental results) that might yield a more revealing interpretation of this experiment.

Chapter 10

Uses of Covariance Analysis

Students sometimes ask 'What is the difference between correlation, regression and covariance analysis?' The answer is none – in the sense that they all use the same statistical process involving calculation of cross products between pairs of variables.[a] If we are interested in the question whether two variables are related, we speak about *correlation* and focus our attention on the correlation coefficient, r; if we are interested in the dependence of one variable on another, we call this *regression* and describe the relationship with an equation such as $y = a + bx$, where b is the regression coefficient; and if we are interested in adjusting one variable to take account of variation in another variable (that is we wish to calculate $y - b(x - \bar{x})$), we call it a *covariance* analysis. In all three cases, however, we shall calculate the covariance and the regression coefficient and the correlation coefficient.

Covariance analysis is at the heart of epidemiological studies, but we do not have time to deal with that topic in the pages of this book. In formal experimentation where chosen treatments are applied, covariance normally has one of two distinct purposes. If we measure a variable such as body weight or early lactation milk yield *before* applying experimental treatments, we may then seek to improve the *precision* of an experiment by using these *preliminary* data to adjust measurements of response made after the treatments were imposed. If, on the other hand, we record a number of variables *during* the course of a trial, such as feed intake and liveweight gain and the yield of milk or eggs, we may then wish to use covariance analysis to help in the *interpretation* of our experimental results. Although the statistical procedure is the same in both these applications, it is important to be clear about your purposes when embarking on a covariance analysis if you are to avoid serious logical errors in drawing conclusions.

[a] The quantity $\Sigma(x - \bar{x})(y - \bar{y})$ is called a *sum of products* and dividing this by the appropriate degrees of freedom gives the *covariance*, cov_{xy}.

Covariance Adjustment Using Preliminary Variables

Suppose that we conduct an experiment with 16 dairy cows, allocating four cows to each of four treatments. We decide not to use a Latin square because we are interested in responses to treatment as measured over a whole lactation (see Chapter 5 for a discussion of change-over designs and their limitations). It is, however, possible to record milk yield and feed intake for each cow for a short period (perhaps 2 weeks) after calving and before she is allocated to a treatment. This will allow us to make an estimate of the potential yield of each cow for that lactation. Experience has shown that there is often a better correlation between yield in a whole lactation and yield in the first few weeks of that lactation than, for example, between one lactation and the next. It is wise to use some judgement about when exactly to take a preliminary measure of milk production. A cow that has had a difficult calving may take longer to come into full milk and there are individual differences in the rate at which cows recover their normal appetite after calving. Animals can join the experiment when they are ready, rather than on some predetermined date or at a fixed interval from calving. Major differences in calving date are usually controlled by blocking. That is, the first four cows ready to join the experiment would be allocated at random to treatments 1 to 4 and would make up block 1. The next four cows would form block 2, and so on.

Table 10.1 gives the milk yields recorded in such an experiment. Similar data would usually be available for protein and fat content of the milk, for feed intake and for liveweight changes, and might be handled in a similar way.

Analyses of variance of the two sets of data (see Appendix 18) indicate no significant differences due to treatment in either period. You would be amazed if 'treatment effects' were found in the preliminary data but are probably disappointed that nothing is showing in the figures for yield while on treatment. The next step is to draw a graph (Fig. 10.1) examining the relationship between x and y in this experiment.

Table 10.1. Preliminary yields (x, kg day^{-1}), measured in the 2 weeks before treatments were applied, and yields (y, kg day^{-1}) during 36 weeks of treatment for 16 dairy cows allocated to four dietary treatments.

Treatment		Individual yields					Totals	Means
A	x	26.1	20.4	17.5	23.4		87.4	21.85
	y	21.5	17.9	17.6	25.7		82.7	20.68
B	x	21.4	25.6	23.0	16.9		86.9	21.73
	y	18.4	25.1	19.8	15.8		79.1	19.78
C	x	16.4	24.8	19.0	18.5		78.7	19.68
	y	12.2	21.6	18.3	19.1		71.2	17.80
D	x	26.2	16.9	18.7	21.3		83.1	20.78
	y	21.9	12.0	15.1	18.0		67.0	16.75
					Grand totals	x	336.1	
						y	300.0	

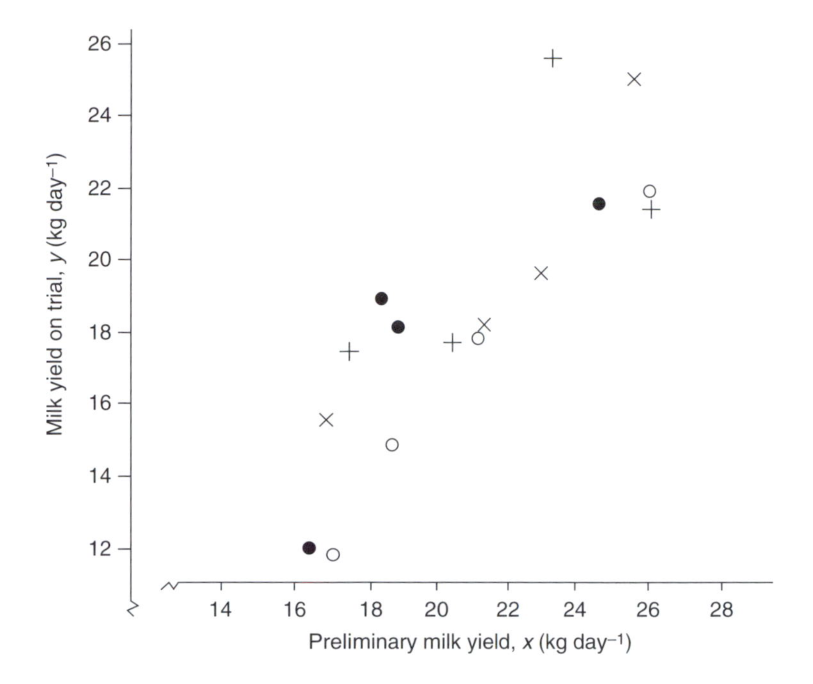

Fig. 10.1. The relationship between preliminary milk yield and yield while on treatment for four dairy cows on each of four treatments: A (+), B (×), C (●) and D (○).

Two important questions must be answered:

1. Is there a relationship between x and y?
2. Is the relationship similar for all the treatments?

We need to have a positive answer to both these questions before proceeding with a routine analysis of covariance. The answer to the first question is pretty obviously 'Yes' in this case. More formally, we could measure the correlation, but asking whether r is significantly different from zero is not particularly helpful. It is possible, for example, to have a correlation coefficient of 0.50 which, with 12 d.f., is not significantly different from zero (at $P = 0.05$) and yet might be a useful correlation for making covariance adjustments to treatment means. All that can be said for certain is that if the correlation is close to zero there is no point in proceeding with a covariance analysis.

The second question is important and is often overlooked. There are formal methods for testing whether the regressions for the separate treatments are parallel, but visual inspection of a graph is often enough to satisfy you that you do not need to bother. Some computer packages will have appropriate tests built in and will issue a warning (or refuse to proceed) if the treatment regressions have significantly different slopes; but not all packages do this and you should be aware of the risks. Lest you should think that non-parallel treatment regressions are unlikely to crop up, you should consider Fig. 10.2.

Failure of the second assumption (that treatment regressions are parallel) is most often found where some treatments place a limit on animal performance while other treatments allow the animals to reach their full potential. If you find yourself with a case where there is some relationship between preliminary and subsequent yields, but the treatment regressions are not parallel, you should seek professional statistical advice. There are ways of dealing with these problems, but each one is apt to be a special case requiring a special solution.

Since there is no suspicion of non-parallel regressions in Fig. 10.1, we can proceed to an analysis of covariance. The arithmetic steps are given in Appendix 18 and the results of the analysis are illustrated in Fig. 10.3.

The yield for each cow is adjusted by calculating:

adjusted yield, $y' = y - b(x - \bar{x})$.

That is, for a cow that was better than the average at the start we knock a bit off and for a cow that was initially below average we add a bit on. In graphical terms, this is equivalent to sliding all the points in Fig. 10.3 up or down the regression slope until they lie on the vertical line marked $\bar{x}$. Notice that the spread of the adjusted values on the y scale is much less than the spread of the

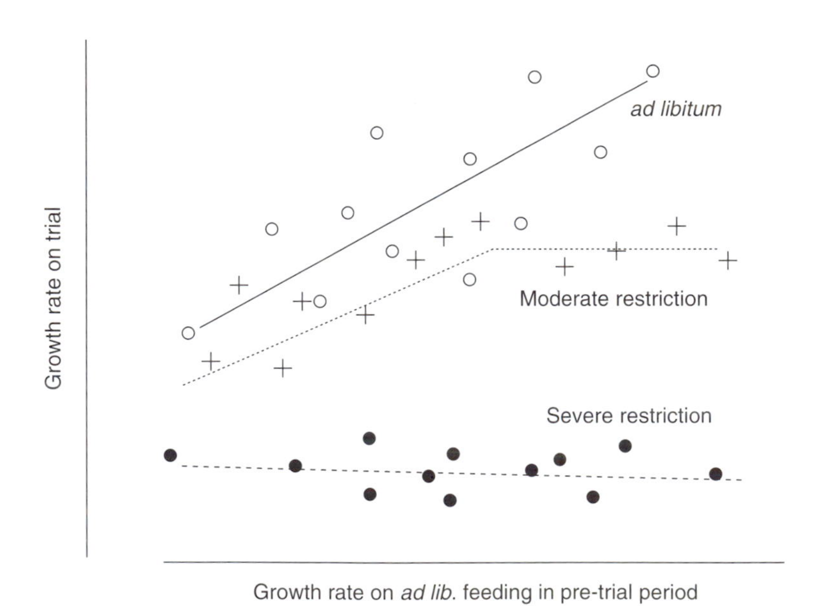

Fig. 10.2. Results of a pig experiment in which growth rate on *ad libitum* feeding was measured for 4 weeks after weaning. Pigs were then allocated to three treatments: *ad libitum* feeding and two levels of controlled feeding (each control-fed pig received a fixed daily allocation of feed in two meals).

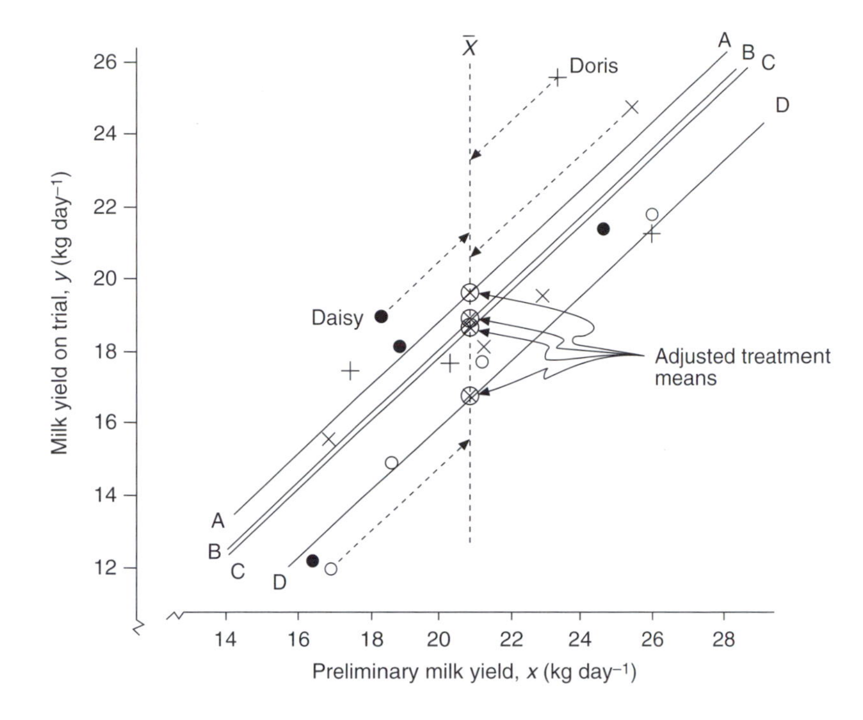

Fig. 10.3. A graphical presentation of the covariance adjustments. Daisy was a below-average cow in the initial period and therefore her yield is adjusted upwards to an estimate of what it would have been if she had been just average. Doris was a higher-yielding cow and has her estimated yield correspondingly reduced. Treatments: A, +; B, ×; C, •; and D, ○. The lines AA, BB, etc. represent the within-treatment regressions fitted with the assumption that these are parallel for the four treatments.

unadjusted values. This is reflected in the error variance before and after adjustment, as shown in Table 10.2.

The error variance is reduced from 15.854 to 4.339 and the CV has fallen from 21% in the unadjusted data to 11% after adjustment. This is a reflection of the very high correlation ($r = 0.8655$) between preliminary milk yield and yield on trial in this experiment. A more typical correlation between milk yield

Table 10.2. Comparison of analyses with and without covariance adjustment.

	Observed yield (y)			Adjusted yield ($y - bx$)		
Source	d.f.	m.s.	F	d.f.	m.s.	F
Treatments	3	12.878	(<1)	3	6.256	1.44 NS
Error	12	15.854		11	4.339	

measured around 2–4 weeks after calving and yield for the following 6–8 months would be between 0.5 and 0.7.

The proportional reduction in error s.s. is r^2. Thus a correlation of 0.5 will reduce error s.s. by 25% and a correlation of 0.8 will reduce it by 64%. The reduction in the mean square is slightly less because we have to pay one d.f. for fitting the linear regression. In the above example,

$$4.339 = (12/11)\{15.854(1 - 0.8655^2)\}$$

The treatment means before and after adjustment are given in Table 10.3. The method of making the adjustment is given in Appendix 18. Treatment C, which had the least favourable set of cows initially (see Table 10.1), has had its relative position improved by the adjustment and treatments A and B, which had good cows initially, have been brought down a little. After adjustment, treatment D is distinguishable from the other three treatments. Although the gap between A and D is not quite significant at $P = 0.05$ it exceeds the 10% probability boundary and so gives the experimenter something to write about. The real moral of this story, of course, is that four cows per treatment are simply not enough, even when the apparent responses are of the order of 17% and the correlation between initial and final yield is as high as 0.87.

Notice that, provided there is a useful correlation, covariance adjustment using preliminary measurements does two things: (i) it gives better estimates of the real responses to treatment by providing adjusted treatment means; and (ii) it reduces the error term.

Multiple Covariance

If you have measured more than one variable before the start of an experiment, for example, initial milk yield and body weight, it is perfectly possible, given access to suitable computing power, to use both variables as a basis for adjusting the data obtained during the experiment. This does not bring in any new principles but does involve the same questions as for a simple covariance analysis: is there a linear correlation between each preliminary variable and the response variable that is to be adjusted, and are the within-treatment regressions

Table 10.3. Treatment means (kg day^{-1}) and their standard errors before and after adjustment by covariance on initial milk yield.

Treatment	Yield (unadjusted)	Adjusted yield
A	20.68	19.91
B	19.78	19.12
C	17.80	19.02
D	16.75	16.96
SEM	±1.99	±1.05

parallel? The adjusted treatment means for a case where three variables x_1, x_2 and x_3, are defined in advance of the experiment (e.g. preliminary milk yield, liveweight and parity) would be calculated as:

$$y' = y - b_1(x_1 - \bar{x}_1) - b_2(x_2 - \bar{x}_2) - b_3(x_3 - \bar{x}_3).$$

Ideally, the three x variables should be independent (i.e. not themselves correlated), but this condition is seldom strictly realized and useful results can be obtained with variables which are only moderately correlated, as would be the case for yield, liveweight and parity.

Blocking versus Covariance

It may have occurred to you that blocking animals on the basis of a preliminary variable also improves the reliability of the treatment means and reduces the error. Indeed, in some cases there will be a choice between using preliminary measurements as a basis for blocking or using them to make covariance adjustments after the experiment is completed, or doing both of these things. Which is the better procedure?

If you expect a linear relationship between the preliminary variable and the experimental measurements, then covariance is more powerful and is to be preferred. It takes account of the exact deviation of the individual animal from the general mean of initial values rather than assigning each animal to a class where only the deviation of the block mean from the general mean is accounted for. Whether this matters does depend upon the distribution of the initial values and, in particular, how much variation there is in those numbers for members of the same block. If, for example, you run a goat experiment in which all the goats assigned to block 1 have an initial weight of 28–29 kg and all the goats in block 8 weigh 36–37 kg there is not much point in bothering with covariance, either as an alternative or as a supplement to blocking. In other cases it is a sensible plan to assign the animals to blocks initially and then to consider doing a covariance analysis at the end if the data suggest that it might be valid and helpful. That covariance analysis must take account of the blocking that has been done. The question that has to be answered is whether there is a regression of y (response) on x (preliminary measure) within treatments *after* the data have been adjusted for block effects.

One argument in favour of blocking on initial data, even when you intend to do a covariance analysis at the end using the same data, is that the treatment means generated during the course of the experiment will be better estimates of the true responses if random variation in the starting material has been partly controlled by blocking. A good experimenter is always anxious to know what is happening in his or her experiment and it is disturbing if effects that looked interesting during the course of the trial disappear after covariance adjustment. This is less likely to happen if major variation present at the beginning has been balanced across treatments by blocking.

In lactation experiments, blocking on initial yield is often not practicable because the cows to be used do not calve sufficiently closely in time. It is possible to wait until all the cows have calved and then to apply treatments to all replicates at the same time; but most experimenters would prefer to assign treatments at a similar stage of lactation, which implies different starting dates for different groups of cows. Here the sensible plan is to use calving date as a basis for blocking and to use preliminary yield for covariance adjustments. Another example where blocking on one variable and covariance on another makes sense is in experiments with young pigs. Litters of pigs, which may be separated in time, can be treated as blocks, while the starting weights of piglets within each litter can be used as a covariate for subsequent adjustment.

In most experiments where preliminary evidence is available which could be used either for blocking or for a later covariance analysis, the sensible procedure is to use blocking at the outset and to do the covariance analysis at the end if the evidence justifies this.

Covariance Adjustment as an Aid to Interpretation

It often happens that during the course of an experiment you will measure a number of variables, each of which could potentially respond to treatment. Feed intake, liveweight gain and body composition are obvious examples. Each trait can be analysed separately, but you may wish to ask the question whether animals on certain treatments grew faster because they ate more feed or, indeed whether they ate more feed because they were growing faster. Now, there is a trap here. Covariance analysis can help to say whether two traits are associated but cannot help to prove which is cause and which is effect (or, indeed, whether neither is 'cause', but both are causally dependent on some other factor which you may or may not have measured). In Appendix 19 you will find a discussion of some experimental procedures that can help you to say whether feed intake is causing, or is dependent on, differences in growth rate. However, covariance analysis does enable you to ask whether, in the light of the general association between feed intake and growth rate observed amongst animals receiving the same treatment, the differences in growth rate between treatments are what you might expect, after taking account of the treatment differences in feed intake. This is really asking a question about the efficiency of utilization of different feeds, but asking it in a much more subtle (and useful) way than by calculating the ratio of feed intake to gain.

Because growing animals need a certain amount of feed for maintenance, we should expect that a treatment that leads to higher feed intake would give a greater liveweight gain and a better feed conversion efficiency, without this implying any necessary change in the *net* efficiency of utilization of feed *above maintenance* (see Fig. 10.4). Calculation of liveweight gain adjusted for differences in feed intake allows you to say whether you would have expected any differences in gain between the treatments if all animals had eaten the same amount of feed.

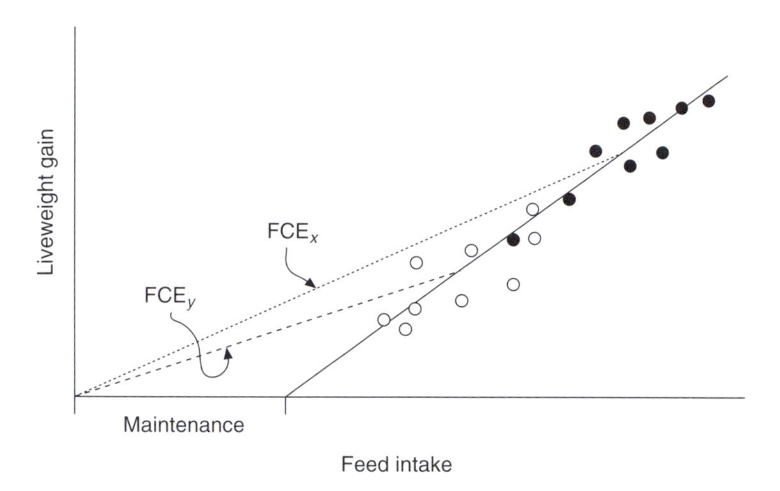

Fig. 10.4. Illustration of the expected improvement in feed conversion efficiency (FCE = gain per unit of feed) which follows from an increase in feed intake. The comparison is between two treatments, *x* (●) and *y* (○), each represented by nine animals.

In a similar fashion, data for carcass fatness can be adjusted for differences in liveweight, some of which will be treatment effects whereas others will be normal variation amongst animals. This will enable you to answer the question 'What would have been the difference between treatments in carcass fatness if all the animals had been slaughtered at the same liveweight?' This would be a relevant question if all animals were killed at the same time (and therefore at different weights). However, you would get a clearer answer to your question by killing batches of animals at two (or more) ages (better still, two or more predetermined weights) because then you will find the answer by interpolation between two treatment means, rather than by extrapolation.

Another example where covariance adjustment should be used more often concerns the adjustment of organ weights to correct for differences in body weight. This correction is often made by expressing organ weight as a fraction of body weight, but this can lead to false conclusions about treatment effects. Consider the data in Table 10.4. The drug administered to the rats clearly reduced body weight and also liver weight, but weight of liver expressed as a fraction of body weight appears to have increased, at least with the higher dose of drug.

An analysis of covariance in this case yields adjusted liver weights (that is, liver weight corrected to a common body weight) which are not significantly different. The explanation of these results is shown in Fig. 10.5. Within the range of data from this trial there is a good linear regression of liver weight on body weight. The regressions are similar for the three treatments (they are drawn independently in Fig. 10.5 although, of course, a covariance analysis fits parallel regressions) and all have positive intercepts. It might surprise you that the linear

Table 10.4. Results of a trial in which a drug was administered to weanling rats at doses of 0, 100 and 200 mg kg^{-1} liveweight. At the end of the experiment, the rats were killed and their body weights and liver weights were recorded ($n = 8$ for each treatment).

	Dose of drug (mg $(kg\ W)^{-1}$)			
	0	100	200	SEM
Body weight (W) (g)	149.7^{a}	131.0^{b}	107.5^{c}	±3.56
Liver weight (g)	7.95^{a}	7.21^{b}	6.20^{c}	±0.204
Liver as % of W	5.35^{a}	5.51^{ab}	5.77^{b}	±0.084
Liver weight adjusted to mean body weight	6.92	7.13	7.31	±0.155

Means in the same row which have different superscripts differ significantly ($P < 0.05$).

equations predict that a rat with zero body weight has a positive liver weight, but it should not. The reason, of course is that the true relationship, drawn from an early embryonic stage to adulthood, is curvilinear and probably allometric. Weight of a component part of the body (C) and body weight (W) usually increase together in a manner described by the equation:

$$C = aW^{k}.$$

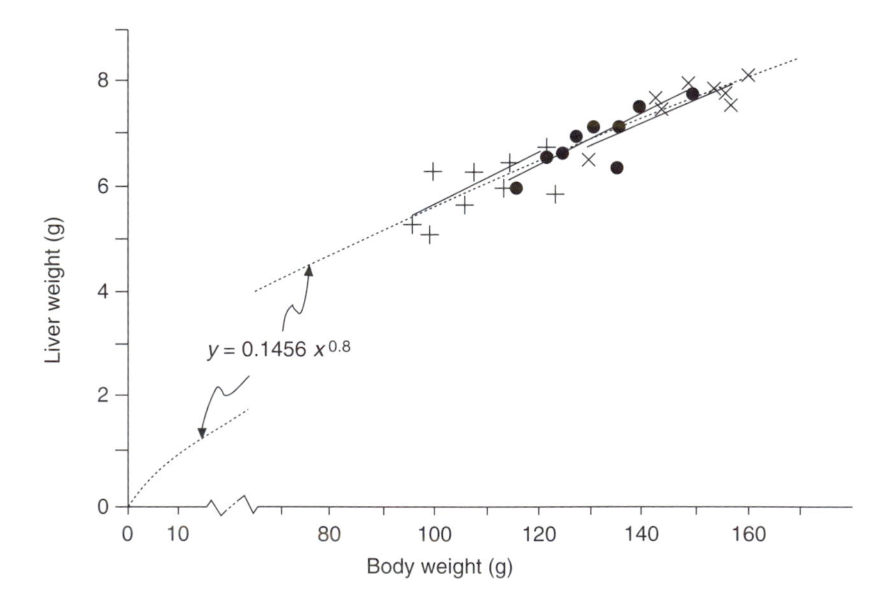

Fig. 10.5. Independent linear regressions fitted to the rat data summarized in Table 10.4. All three regressions have significant positive intercepts (though not as large as appears at first glance: note the piece missing in the horizontal scale). The dashed line indicates an appropriate curvilinear regression which does predict zero liver weight at zero body weight.

If $k < 1$, as it normally is for liver weight, the curve relating liver weight to body weight decelerates with age and large rats will have proportionally smaller livers than small rats.

Whenever you are thinking of taking a ratio between two traits (feed/gain; fat weight/muscle weight; milk yield/body weight) stop and consider what *overall* relationship you would expect for those two traits. If it is not a straight line *through the origin*, you should be using covariance rather than the ratio. Although most of these relationships are curvilinear when you assemble a wide enough range of data, they are all approximately linear for data spanning a narrower range. The advantage of the covariance analysis is that it estimates the average within-treatment regression which applies to your data and (if you have studied a graph or used a suitable computer package) warns you when that relationship is not similar for the several treatments.

Summary

1. Covariance analysis can be employed *either* to *improve precision*, using measurements taken before treatments were applied, *or* to *help interpretation*, involving two or more variables that may have been influenced by treatment.
2. Routine analysis of covariance involves an assumption that the *within-treatment regressions* are *linear* and *parallel*. These assumptions need checking.
3. Covariance using a *preliminary* variable *reduces error* s.s. by a factor of $1 - r^2$, where r is the correlation between preliminary variable and final measurement. Covariance analysis also leads to *adjusted treatment means*.
4. Covariance analysis reveals the associations between variables, but *cannot* be used to *prove* which is *cause and* which *effect*.
5. Covariance analysis is usually more helpful to understanding results than the analysis of a *ratio* between two traits. In particular, *adjustment* for *body weight* should be done by covariance, rather than by dividing by body weight.

Exercise 10.1

Table 10.5 gives data from an experiment in which pigs were fed three diets containing graded levels of jack beans (*Canavalia* sp.). The beans were treated by chopping and then boiling them in water in an attempt to remove soluble and heat-labile toxins. The treated beans were then sun-dried and ground to a meal. Diets were formulated to have equal energy, mineral, vitamin and lysine contents, with jack beans displacing mainly soybean meal. Eight individually fed pigs were allocated to each diet. Each animal was introduced to its assigned treatment when it reached 20 ± 1 kg and fed *ad libitum* until its weight reached 40 kg.

Table 10.5. Feed intake (g day^{-1}) and weight gain (g day^{-1}) for eight pigs on each of three diets containing 0, 80 or 160 g kg^{-1} of a meal made from chopped and boiled jack beans.

	Jack beans in the diet (g kg^{-1})					
	0		80		160	
	Feed	Gain	Feed	Gain	Feed	Gain
	2624	1121	2835	1024	2409	825
	3017	1105	2300	911	2503	892
	2780	1162	2908	1106	2570	1044
	2830	974	2954	1043	2826	1101
	2518	1027	2583	895	2176	816
	2442	978	2998	1222	2794	991
	3281	1337	2477	1071	2204	790
	3295	1272	2219	952	2251	897
Means	2848.4	1122.0	2659.2	1028.0	2466.6	919.5
SEM	105.0	42.0	105.0	42.0	105.0	42.0

1. Perform analyses of variance and covariance on these data to see whether they support an argument that the reduction in growth rate associated with feeding the treated beans can be attributed to the reduction in feed intake observed.

2. Can you think of an experiment that would allow you to say more directly whether the lower feed intake was the cause of depressed growth rate or whether lower feed intake should be regarded as a consequence of reduced growth rate?

Chapter 11

Unbalanced Designs

Most designs used in animal experiments are *balanced* (or orthogonal) in the sense that equal numbers of replicates are used for each treatment and, if there are blocks or other features (e.g. rows and columns in a Latin square), each treatment is equally represented in each block. This balancing is *efficient* in the sense that it gives the greatest possible precision, for a given set of resources, provided the assumptions underlying a routine ANOVA (homogeneity of variance, etc.) are true.

If you consider the allocation of 20 animals to two treatments (say control and injected), the smallest possible SED is obtained by allocating ten animals to each treatment. Using five animals as controls and injecting 15 is expected to result in a larger SED and so a less sensitive experiment, unless you have good reason to think that the injected animals will give *more variable* results than the controls.

There are circumstances, apart from unequal variances, in which you might sensibly choose *unequal replication* of your treatments, without producing an unbalanced design. As mentioned in Chapter 9, you may decide to have fewer replications of the lowest doses in a dose–response experiment, either because those treatments involve serious economic loss or because they cause distress to the animals. Equally, if you were testing a new vaccine to control a virulent disease, you might decide that you needed substantial numbers of animals in two groups to be '*vaccinated and challenged*' and '*unvaccinated and unchallenged*' so as to detect differences in performance between these two treatments, which could be quite small: but you might use many fewer animals in a third group to be '*unvaccinated and challenged*' since the purpose here is only to demonstrate that your culture of the disease organism is alive and kicking and that you have used an effective procedure for transmitting the infection.

These examples of deliberately chosen unequal replication can be incorporated in *balanced designs* in which, if there are blocks at all, the *proportions* (though not the numbers) of animals from the several treatments are the same in all blocks.

If there are no blocks or other features to consider in the ANOVA beyond the treatments, then analysis of unequally replicated treatments presents no problems. If for some reason we have 8, 10 and 5 animals respectively allocated to (or surviving from) three treatments, the ANOVA will show 2 d.f. for the variation amongst treatments and 20 d.f. for a pooled error. Exercise 11.1 gives a worked example of such an experiment and should cause you no difficulty. The SED for comparing each pair of treatments has to be computed separately, but that is very little trouble (Appendix 2 gives the general formula and Appendix 20 gives a numerical example).

Missing Plots

If the experiment described above had originally been designed with ten replicates of three treatments in an RCB design and the data came from 23 survivors of the original 30 animals (see Exercise 11.2), the ANOVA is not straightforward. Block effects will be *confounded* with treatment effects in the block and treatment totals and you cannot derive valid estimates of blocks s.s. or treatment s.s. simply by squaring the 'relevant' totals.

Where only a single result is missing from an otherwise balanced design, one solution is to insert a *missing plot value* in the gap. The formulae for estimating missing plot values in an RCB or LS design are given in older statistical textbooks such as Snedecor (1956). Their purpose is to enable you to put a number in the empty space so that you can do a routine ANOVA using a pocket calculator. If you do not have access to a computer with a statistics package, you will find the formulae useful, but tedious if you need to fill in more than one missing value. However, computer packages now readily provide an ANOVA and best estimates of treatment means for designs with one or more missing values and this saves a great deal of time and trouble. The principles involved in analysing formal designs with missing values are the same as those described below for unbalanced designs.

Unbalanced Designs

There are many reasons why you might choose, or be forced to adopt, an unbalanced design. Some examples arise because an originally balanced design has lost some of its plots; others occur when the natural subdivisions of the material available do not provide you with tidy numbers. For example, you may wish to take account of litter origin in a pig experiment but the litters are of unequal sizes.

In all these cases the underlying assumption is that the data can be represented by an additive model of the kind:

plot yield = a general mean + a treatment effect + a block effect + a random error.

The problem then is to find the set of block means and treatment means (and means for any other effects that were included in the design) which give a minimum value to the residual sum of squares. Computer packages can do this very effectively, provided the imbalance in the design does not lead to total confounding of two effects.

Some Examples

Table 11.1 shows three designs that might have cropped up in a dairy cow experiment. Design (a) – the original plan – is balanced. Although there were half as many Jerseys as Holsteins available, the proportion is the same on each treatment. The breed and diet totals will yield unbiased estimates of breed and diet effects and the interaction is readily estimated. All this can be done with a pocket calculator if you choose.

Design (b) is unbalanced and cannot be readily analysed with a pocket calculator; but a suitable computer package will give you best estimates of the breed and dietary effects, with appropriate standard errors. The breed × diet interaction can be estimated, but you should be cautious in trusting it when the design is as thin as this.

Design (c) is severely unbalanced. No estimate of breed × diet interaction is possible (you need a minimum of 2 × 2 breed × diet combinations to begin to measure interaction). The estimates of diet effects will be highly suspect, since there are no Holsteins on the medium treatment. The Jersey data for the medium treatment can only be tied in by reference to the difference between Holsteins and one Jersey on the high level of feeding. This experiment is now a disaster and it would be best not to attempt to draw any conclusions (except to start out with more animals next time) if you are left with a mess such as this.

A more comfortable position arises when you have lots of pigs available for an experiment, but they come in two sexes and originate from several litters. Figure 11.1 illustrates the material that might be available for a pig experiment involving four treatments.

If we want to balance the experiment for the effects of sex and litter, we shall take the seven blocks of four pigs within the dotted outlines. Notice

Table 11.1. Three different structures (a, b and c) for a dairy cow experiment involving two breeds (Holstein and Jersey) and three feeding levels (low, medium and high). The number of replicate animals for each breed × treatment combination is shown.

	(a)		(b)		(c)	
Diet	Holstein	Jersey	Holstein	Jersey	Holstein	Jersey
Low	4	2	4	1	3	0
Medium	4	2	3	2	0	2
High	4	2	3	1	3	1

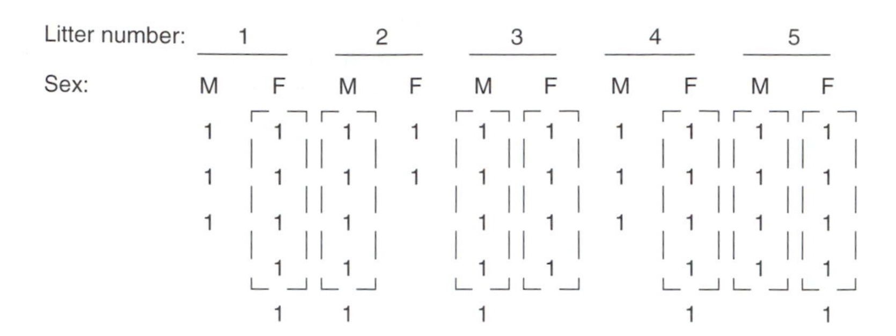

Fig. 11.1. Numbers of pigs available for an experiment by sex and litter origin. The dashed lines mark out seven blocks, each containing four pigs of the same sex and litter origin.

that this will not give very efficient estimates of the separate effects of litter and sex, because litters 1 and 4 are only represented by female pigs and litter 2 only contributes males. That degree of confounding does not, however, reduce the efficiency of blocking. The four treatments will be compared with the mean effects of blocks completely removed and we do not really care how much of the block variance is attributable to sex and how much to litter origin.

However, there are 41 pigs available and it would be a pity to discard 13 of them, just for the sake of achieving balance. Given access to a computer package capable of giving you least-squares estimates of litter, sex and treatment effects, your experiment will be more powerful if you use all of the material. There are various strategies for allocating treatments to the 40 pigs, but a sensible approach would be to start by assigning treatments at random within the sets of four marked off in Fig. 11.1. That leaves you with 13 pigs outside the boxes, eight males and five females. Allocate the four treatments at random to the eight males with the restrictions that each treatment is used twice but no treatment appears twice within the spare male pigs of one litter. Follow the same principles for the females, allowing one of the four treatments (whichever comes out of the bag) to be allocated twice, but not within the same litter.

Examples requiring the allocation of treatments in a fashion that is not perfectly balanced occur quite commonly in animal trials. The usual reasons are that you are faced with different breeds or strains, different sexes, different parities, different litter origins (pigs) or different litter sizes (sheep). The general rule is to achieve the most balanced distribution of treatments that is feasible and not to discard substantial amounts of potential experimental material for the sake of balancing the design.

In all cases where you are faced with an imbalanced design, consult a statistician before you start. Some designs can be analysed to recover useful information but some cannot.

Summary

1. Unbalanced designs can arise by *accident* or by *design*.
2. *Least squares analysis* can usually provide unbiased estimates of treatment effects, but there are some limitations. If the holes in the design are too great, no amount of fancy analysis will recover useful information.
3. Consult a statistician before deliberately embarking on an experiment with an unbalanced design.

Reference

Snedecor, G.W. (1956) *Statistical Methods*. Iowa State Press, Ames, Iowa.

Exercise 11.1

Analyse the data in Table 11.2 and present a table or diagram showing SEDs, derived from a pooled variance, for each pair of treatment means.

Table 11.2. Mean liveweight change (g day^{-1}) for beef cows on range receiving two supplementary feed treatments throughout the dry season and for controls receiving no supplement.

	Supplement 1	Supplement 2	Unsupplemented
	+112	−9	−276
	+17	+121	−178
	−22	+24	−139
	+144	+17	−319
	−17	−48	−212
	+206	+92	
	+63	+40	
	+114	+97	
		−17	
		+63	
$n =$	8	10	5
$\Sigma x =$	+617	+380	−1124

Exercise 11.2

The animals referred to in Exercise 11.1 were the survivors from an experiment in which 30 animals were initially assigned to ten blocks on the basis of a combination of liveweight and condition score. Three treatments were allocated at random within blocks. The full structure of the experiment is shown in Table 11.3.

Table 11.3. Mean liveweight change (g day^{-1}) for surviving beef cows, originally allocated to ten blocks, and then placed on range during the dry season with no supplementary feed or with one of two supplementary feed treatments.

Block	Supplement 1	Supplement 2	Unsupplemented
1	+112	−9	−276
2	+17	+121	−178
3	−22	+24	
4	+144	+17	−139
5		−48	−319
6	−17	+92	
7		+40	
8	+206	+97	
9	+63	−17	−212
10	+114	+63	

Find a suitable computer package and analyse these data so as to recover the block variance. Compare the error you obtain with that found in Exercise 11.1.

Chapter 12

Repeated Measures

This is a large and complex subject and we shall only nibble at the edges of it. A proper treatment of the problems that arise with repeated measures can be found in a book by Crowder and Hand (1990). The purpose of this chapter is to make you aware of a potential problem and to indicate some simple ways of avoiding trouble.

A statistical difficulty arises when you measure the same animals repeatedly, as happens in a lactation trial where you may collect feed intake and milk output data on a daily basis; or in a growth trial when you weigh animals at weekly intervals; or in a physiological experiment where you may make observations on an hourly basis.

The trouble is that adjacent measurements are likely to be *correlated* and are thus *not independent* estimates of the performance of the animal.

Consider the matrix of results represented in Fig. 12.1. Here you have 12 animals allocated to three treatments with data reported on a weekly basis for eight consecutive weeks. It might seem obvious that you should perform an ANOVA of the kind outlined in Table 12.1; but *this is wrong*. An ANOVA with 69 d.f. for error is *cheating*. It implies that each treatment was replicated 32 times which is not the case. Repeated measurement of the same cow does not amount to replication because cows cannot be independently randomized to treatments for each week's data collection (if you think they could or should be, read Chapter 5 on change-over designs, which may make you think again). In the present case, cows were allocated *once* to treatments at the beginning of the trial. If, by chance, you put your best cow on treatment 3 and your worst cow on treatment 2, that random effect will probably persist throughout the experiment. Looking at the weekly yields might then suggest that there was a *consistent* advantage of treatment 3 over treatment 2, but that apparent consistency would be the result of repeatedly measuring the same cows. If, on the other hand, you look at the total yields for the 8-week trial, you might see no consistent superiority of one treatment over another, because now you are *judging differences* between treatments *in the light of variation seen amongst animals receiving the same treatment*. This, as we have remarked before, is the only right way to reach sound conclusions.

Weeks → Treatment	1	2	3	4	5	6	7	8	Rep. totals	Treatment totals
A rep. 1	•	•	•	•	•	•	•	•	●	
rep. 2	•	•	•	•	•	•	•	•	●	
rep. 3	•	•	•	•	•	•	•	•	●	
rep. 4	•	•	•	•	•	•	•	•	●	●
B rep. 1	•	•	•	•	•	•	•	•	●	
rep. 2	•	•	•	•	•	•	•	•	●	
rep. 3	•	•	•	•	•	•	•	•	●	
rep. 4	•	•	•	•	•	•	•	•	●	●
C rep. 1	•	•	•	•	•	•	•	•	●	
rep. 2	•	•	•	•	•	•	•	•	●	
rep. 3	•	•	•	•	•	•	•	•	●	
rep. 4	•	•	•	•	•	•	•	•	●	●
Week totals	○	○	○	○	○	○	○	○		

Fig. 12.1. An array representing the results of a lactation experiment conducted over 8 weeks with 12 cows allocated to three treatments in an RCB design. rep = Block.

However, an ANOVA of total (or mean) yields for the whole experiment involves no difficulty due to repeated measures. Nor does a separate analysis of each week's data. Each of those analyses would be based on 11 d.f. (6 d.f. for error) as shown in the top half of Table 12.2.

There is still a technical difficulty with the bottom half of Table 12.2, because weeks cannot be allocated at random, they come in ready-made consecutive strips! For that reason, the interpretation of what might be called a 'sampling variance' (42 d.f. in this example) is a little dodgy. It should not be taken as a reliable estimate of the extent to which *independent* estimates of the

Table 12.1. An improper analysis of the experiment depicted in Fig. 12.1.

Source	d.f.
Replicates	3
Treatments	2
Weeks	7
Treatments × weeks	14
Residual	69
Total	95

Table 12.2. A better ANOVA for the experiment depicted in Fig. 12.1.

Source	d.f.
Replicates	3
Treatments	2
Error	6
Cows	11
Weeks	7
Replicates × weeks	21
Treatments × weeks	14
Replicates × treatments × weeks	42
Total	95

yield (or the butterfat or whatever else you have measured) of one cow may vary. If this puzzles you, think about sampling the same cow for milk protein content on 7 consecutive days or on 7 days at monthly intervals. You would surely expect different variances for these two exercises? That is because measurements taken at long intervals are apt to be less well correlated than measurements closely related in time.

However, Table 12.2 is a distinct improvement on Table 12.1 and is less likely to lead to false conclusions. In particular, the error with 6 d.f. will give you the appropriate SEM for judging treatment means.

Time Trends

Sometimes you may be particularly interested in the treatments × weeks component of variance. That is, you want to ask whether the pattern of response to treatment varies with time. That would certainly be a relevant question in a growth trial and might be important in a lactation experiment. One of the best ways of handling this problem is to *fit a line* (or a curve) to the measures repeated over time for each animal and to compare the parameters of the lines. Thus, for animals gaining weight steadily during a trial it is often possible to represent the data for each animal with a straight line. This will not work if there are changes of regime, such as occur on turning animals out to pasture at the end of a winter feeding period or during a prolonged tropical dry season, but it works in many other short-term trials. The slopes of linear regressions fitted separately to the weighings for each animal give the best estimates of mean growth rate for each animal. All the weighings are used to obtain that estimate, rather than just the initial and final ones, and if treatments diverge with time this will be reflected in the regression slopes. The regression coefficients can be subjected to a standard ANOVA which will allow you to judge whether the slopes differ significantly between treatments. You will find an example of this procedure in Exercise 12.1

If the growth curves or lactation curves you have to analyse are definitely *not linear*, you can still estimate the response for each animal by fitting quadratic, logarithmic or exponential equations. Now, however, the trend for each individual is represented by two numbers and you cannot readily subject these pairs of numbers to a routine ANOVA. Moreover, the two parameters will be highly correlated and you should therefore not attempt a multivariate analysis. However, you can use the fitted equations to estimate the weight change over a chosen interval of time or the time taken by each animal to reach a given weight. These interpolated weight changes or predicted times can then be analysed for treatment effects in the usual way.

Weighing Ruminant Animals

This may be a good moment to point out that errors due to variable *gut-fill* in ruminants can represent large quantities in relation to liveweight changes that you wish to detect. This is usually not a serious difficulty with young animals, which may double their size during the course of an experiment; but, with adults, changes in liveweight, which mainly reflect changes in stored body fat, are often important to an understanding of the biology of your problem and yet may be difficult to detect against a background of variable gut-fill. Moreover, the changes might not be unidirectional. A good cow will lose weight in early lactation and then put most of it back in the early stages of the subsequent pregnancy. These changes can be of the order of 20–40 kg, but a cow can also lose 20 kg of milk at one milking and she can easily gain 30 kg by eating and drinking between one milking and the next.

Scoring the condition of the animal is only a partial solution to your difficulties, because *condition scoring* is subjective, not highly repeatable and thus not able to detect small changes reliably. There is also a special problem with condition scoring your own animals. You probably know which treatment each animal is receiving and are therefore wide open to making unintentionally biased judgements. If you decide to use condition scoring, it is important that the same person does the scoring every time and that he or she does not know which cow is on which treatment.

The first and obvious rule when weighing animals at regular intervals is always to do it at the same time of day. It is also important to ensure that the routine of milking, feeding and handling animals is as regular as possible. Secondly, if you have taken a sequence of weights covering a short period, estimate weight change by regression, not by subtracting initial weight from final weight. Thirdly, if you need good estimates of weight change over periods of a few weeks, consider weighing the animals repeatedly at the beginning and end of each period (e.g. on three consecutive days) as an alternative to weighing them at regular intervals throughout the period. This will certainly help with change-over trials where the treatment periods may last 3–6 weeks.

Animals of all sorts readily become accustomed to a routine and, as well as being good for their general welfare, it will help you to gather reliable data if you weigh your animals, whether they be pigs, chickens or ruminants, quietly and gently and following the same routine at every weighing.

Automatic weighing systems have been devised for cows, pigs and chickens, but a discussion of the reliability of these gadgets will not be attempted here, since technology moves rapidly and comments written today may be out of date by the time this book is published.

Summary

1. Repeated measurements must *not* be analysed as though they represented additional *replications* of treatments.
2. With a series of weighings taken over time, it is permissible to analyse mean weight or overall gain, but dangerous to try and analyse *weight* × *time*.
3. A good method for handling a series of weighings is to fit a line or a curve to the results obtained for each animal.
4. With straight-line responses, an *analysis of the regression coefficients* is the best way of handling time-series data.
5. If the time trends are curvilinear, fitted curves can be used to estimate *weight gain* over a *given interval of time* or the *time taken* to reach a *given weight*.
6. Estimating short-term weight changes in adult ruminants is difficult, because of variations in *gut-fill*.

Reference

Crowder, M.J. and Hand, D.J. (1990) *Analysis of Repeated Measures*. Chapman & Hall, London.

Exercise 12.1

Table 12.3 gives the liveweights of 24 Simmental × Friesian heifers weighed at fortnightly intervals in an experiment testing diets based on maize silages, made at different dates, which varied in their dry matter content and nutritive value (1 = low, 2 = medium, 3 = high dry-matter silage). The animals were placed in three pens, grouped according to their liveweight, just before the experiment started. Calan gates were used to allow animals within pens to be fed separately and their individual food intake was recorded. The three treatments were each randomly assigned to eight animals.

1. What do you think are the chances that the weight of heifer number 19 in week 4 is a recording error? What will you do about it?
2. What is wrong with the ANOVA framework in Table 12.4 for analysing these data?

Table 12.3. Liveweights (kg) of 24 individually fed heifers receiving three treatments in three pens.

Pen	Animal	Treatment	Week					
			0	2	4	6	8	10
1	1	3	368	380	377	389	413	426
	2	2	367	386	411	411	435	444
	3	1	377	398	421	426	441	458
	4	2	371	388	391	407	416	427
	5	2	353	384	398	406	424	445
	6	1	364	383	388	401	420	432
	7	3	371	381	396	398	414	415
	8	1	372	389	402	398	435	441
2	9	3	363	378	396	406	429	447
	10	1	383	411	418	426	453	461
	11	2	371	400	410	426	443	453
	12	3	375	398	410	418	450	459
	13	3	374	388	399	414	424	438
	14	3	385	399	405	422	442	456
	15	1	384	410	423	440	462	475
	16	1	373	392	411	417	437	455
3	17	3	394	404	426	430	445	461
	18	2	390	411	423	419	445	467
	19	2	389	413	440	435	464	472
	20	1	381	406	422	432	447	463
	21	3	384	404	422	433	457	475
	22	2	401	427	454	451	468	489
	23	1	391	406	433	444	455	462
	24	2	374	402	420	424	444	455
Weekly means			377.3	397.4	412.3	419.7	440.1	453.2

3. Estimate weight gain by regression for each animal and perform an ANOVA of the estimates that you obtain.

4. Is the precision of estimating weight gain by regression in these data any better than the precision obtained by taking the difference between finishing and starting weights?

Table 12.4. Proposed ANOVA for analysing the data in Table 12.3.

Source	d.f.
Weeks	5
Treatments	2
Weeks × treatments	10
Error	126
Total	143

Chapter 13

Discrete Data

In Chapter 9 we listed the assumptions that underlie routine analysis of variance. The first and most important of these is that your data should be drawn from a continuous variable which is normally distributed. However, it is not uncommon to encounter data that are *not continuous*, either because the results are qualitative, not quantitative (e.g male or female, alive or dead, pregnant or not pregnant), or because the numbers are small (e.g. litter size in sheep or goats). You cannot be 'a little bit pregnant' and a single ewe cannot have a litter of 1.7 lambs.

If you happen to have several *large groups* of animals, then the numbers become, for all practical purposes, continuous and approximately normal, even though, at its root, the character is discontinuous. A herd of cows can have a conception rate (to single insemination) of 53.7% and a flock of sheep can have a mean litter size of 1.73. Comparisons amongst numerous herds or numerous flocks can thus be made by treating the data as continuous. But in formal experiments employing large animals, it is not usually possible to allocate replicated *groups* to each treatment; the basis of replication is almost always the *individual* animal. Some of the data collected may then be *categorical*, meaning that the result for any one animal falls into one or another of a small number of categories. Although litter size in cattle, sheep or goats can be represented by a number, you can also think of it as a set of categories (singles, twins or triplets). 'Alive or dead' are clearly two mutually exclusive categories which cannot be realistically represented by numbers, even though you might assign dummy values of 0 and 1 to these conditions for certain analytical purposes.

For pigs and rabbits, where the litters are larger, it is usually safe to treat litter size as though it were a continuous variable, and the same goes for egg numbers at a single ovulation in polytocous mammals or for egg laying by poultry over an extended period. Individual litter sizes in pigs might range from 6 to 14 and the number of eggs laid by one chicken in a month might range from 20 to 31, and such data will generate variances that can be treated as part of a normal distribution. However, if you were considering eggs laid by individual hens on a single day, that would be a discrete variable with values limited to 0, 1 or 2.

Behavioural analysis also commonly yields categorical data. For example, you might observe times spent grazing, resting, ruminating, etc. for groups of cattle in paddocks with different grazing management policies. Although time is a continuous variable, the total of the times that you record is fixed by your choice of observation periods (if cows take more time ruminating on a given treatment, they necessarily spend less time doing other things) and so you are analysing a *pattern* of results within a fixed overall framework. You therefore have categorical data.

The usual method of analysing categorical data is to employ a *chi-squared test* (χ^2). We will assume that you know about this test but, if you need a reminder, there is a worked example in Table 13.1. A table giving probabilities for χ^2 values with specified d.f. will be found in Appendix 24. Although you have probably used χ^2 before, there are some traps for the unwary which are worth pointing out here.

Snags with the χ^2 Test

A common mistake made by beginners is to use derived proportions instead of the original numbers from which the proportions were calculated. That is wrong. If I tell you that in two groups the mortality rates were 6% and 14% you might wisely enquire how large the samples were. If I then tell you that one animal out of 16 died in the first group but two out of 14 in the second, you will be unimpressed. If, on the other hand, I tell you that group A consisted of 217 animals, 13 of which died and group B initially had 221 animals, of which 31 died, you may think that the difference between 13 and 31 deaths is compelling evidence. You would be right. The odds against this being a chance difference are more than 100:1. It is the *numbers* that matter, *not the proportions*. Moreover, the numbers of animals that survived contribute to the story and to the χ^2 test, as shown in Table 13.1.

Another complication with a χ^2 test is that the results are unreliable if you have very small numbers in any cell of your table of results. Zero and one are

Table 13.1. A chi-squared analysis of deaths and survivals in two groups of animals.

	Observed numbers (O)			Expected numbers (E)		Deviations ($O - E$)		$\chi^2 = (O - E)^2/E$	
Group	A	B	Totals	A	B	A	B	A	B
Died	13	31	44	21.8	22.2	−8.8	8.8	3.55	3.49
Survived	204	190	394	195.2	198.8	8.8	−8.8	0.40	0.39
Totals	217	221	438	217	221	0	0	$\Sigma\chi^2 = 7.83$	

d.f. for an $r \times c$ table (where r = number of rows, c = number of columns) $= (r - 1) \times (c - 1) = 1$ in this case: therefore $P < 0.01$.

decidedly dangerous and numbers up to about five are suspicious, meaning that probabilities estimated when you have such numbers amongst your results will not be reliable. One recommended method of getting round the problem of small numbers in some cells is to amalgamate categories. When you have several categories in your results, such as a 2 × 5 table, which would arise if you were to compare the numbers of goats on two treatments giving birth to 0, 1, 2, 3 or 4 kids, the problem of very small numbers in the fringe categories may be resolved by lumping some groups together. Thus, if you have very few goats with no live kids and few with quadruplets, dropping the 0 group and amalgamating the 3 and 4 categories will give you a 2 × 3 table, showing the number of goats that delivered 1, 2 or '3 or more' kids. If amalgamation of groups does not solve your problem, perhaps because you only have two categories for each treatment, run the χ^2 test but report the result as an approximate estimate of probability.

Estimating the Expected Outcome

In Table 13.1 the expected numbers dying in each group are readily derived from the null hypothesis that there is no difference (other than chance) between the groups and therefore the expected proportion of deaths is what we observe in the entire sample, i.e. 44/438. Sometimes things are a little more complicated than this and expectations must be derived with the help of a little binomial theory.

Suppose that we record the frequency of twinning in a population of cows and get the results shown in Table 13.2. We might ask the question whether there is any evidence in these data that cows that have had twins at their first calving are more likely than those bearing singles at the first calving to produce twins at the second parity. In other words, is there evidence that twinning is to some extent a repeatable phenomenon? Notice that our null hypothesis does not include the proposition that the rate of twinning is constant for successive parities. We can ask the question about repeatability while accepting that older cows are more prone to have twins.

The rate of twinning for first calvings in these data is 56/10728 = 0.0052. Similarly, the overall rate of twinning for the second parity is 153/10728 = 0.0143. Expected results after two parities are then as set out in Table 13.2. The use of binomial notation to find appropriate expectations is particularly helpful if you wish to pursue the question of repeatability for another three parities, as Bowman and Hendy (1970) did in their original report of these data. The problem is slightly more complicated than the one presented here, as not only does the twinning rate increase with age, but the number of records available declines with age and the cows contributing data are not the same ones from one parity to the next.

There is 1 d.f. for the sum of chi-squares in Table 13.2 because, although we have set out four categories in a line, this is really a 2 × 2 table (two parities × two birth types).

Table 13.2. Twinning records at the first two calvings for British Friesian cattle in southern England.

First parity n	Cows with twins	Cows with singles
10728	56	10672

$p_1 = 56/10728 = 0.0052$; $q_1 = 10672/10728 = 0.9948$.

Second parity	Cows with no twins	Cows with a single followed by twins	Cows with twins followed by single	Cows with two pairs of twins	n
Observed (O)	10524	148	51	5	10728
$p_2 = (148+5)/10728 = 0.0143$; $q_2 = (10524+51)/10\,728 = 0.9857$.					
Expected	$n.q_1.q_2$	$n.q_1.p_2$	$n.p_1.q_2$	$n.p_1.p_2$	
$E =$	10519.6	152.6	55.0	0.8	10728
$\chi^2 = (O - E)^2/E$	0.002	0.139	0.291	22.050	
$\Sigma\chi^2 = 22.482$ with 1 d.f., $P < 0.001$					

The expected number of cows with two pairs of twins is very small and therefore we must treat the estimate of χ^2 as an approximation, but the result of the test leaves no room for doubt that there are more cows in this category than we would expect from a null hypothesis that twinning in the second parity is independent of birth type at the first parity. Cows that have had twins once are more likely to have twins again than cows that have never given birth to twins.

Summary

1. Some results involve numbers that are not part of a continuous distribution. Common examples are results classified into *categories* (e.g. alive or dead) and data with *small numbers* (e.g. litter size in sheep).
2. Such data can usually be analysed with the help of a χ^2 *test*. This is easy to apply, but be aware that the test is *approximate* if the number in any one category is *less than 5*.
3. Characters that are inherently *discontinuous* (e.g. number of offspring at one parity) can be treated as continuous variables if analysing results for replicated groups of animals (e.g. mean litter size), because *group means* show continuous variation and are usually approximately normally distributed.

Reference

Bowman, J.C. and Hendy, C.R.C. (1970) The incidence, repeatability and effect on dam performance of twinning in British Friesian cattle. *Animal Production* **12**, 55–62.

Chapter 14

Multiple Experiments

The title of this chapter is used to introduce discussion of two separate procedures which, nevertheless, have something in common. First, you may be *planning* a series of experiments, all bearing on the same question, with the assumption that a series of trials will provide better answers than one trial conducted at one location at one time. Secondly, you may be *reviewing* a series of experiments conducted by yourself and others with the purpose of deriving a general (or average) answer to a question or series of questions.

A graduate student may well be engaged in both of these activities at different times, although it is sad that reviewing the literature, at least in the quantitative and analytical manner discussed below, usually does not happen until after the student's own experiments are completed. It is also sad that the first experiment carried out by the student is so often seen as the definitive trial to answer the question once and for all, rather than as the first in a series of trials which, cumulatively, will lead nearer and nearer to 'the truth'.

Planning Multi-location Experiments

Trials involving the same (or similar) treatments may be replicated in space or in time (or both). Multi-location trials are used routinely in crop experimentation to provide data for variety and treatment effects (and their interactions) covering different soil types and climatic and disease environments. Moreover, such trials are usually repeated for two or three seasons before any pronouncement is made about the relative merits of new varieties or new crop protection products. This is because experience has shown that the effects of weather and disease can alter the ranking of treatments quite markedly and thus claims based on a trial at one location in 1 year have very little value. Their value is, in fact, to show that further testing is worthwhile. It is important to realize that, when a multi-location crop trial is planned this is usually the sequel to a series of preliminary trials conducted in more restricted circumstances, such as the plant breeder's own test station.

With animal experiments it is much less common to plan repetitions of a trial at multiple locations. There are good arguments in favour of repetition, in that more genotypes, more physical and disease environments and more management practices will be sampled in a multi-location trial, and all these factors could have a bearing on the results. But animal experiments are costly to run and there is some justification for the (unspoken) assumption that results obtained at one location in one year can readily be duplicated (if we care to try) at other stations in different parts of the world in subsequent years. The repetition in space and time, which the agronomist regards as an essential feature of his research, is thus normally achieved in the animal world by the independent activity of different research workers.

Nevertheless there are some examples of multi-location testing with animals, and their advantages and disadvantages are worth considering. A multi-location pig trial reported by Braude *et al.* (1962) provides one example. In 1960 it was known that incorporation of an antibiotic in a pig or poultry diet generally improved growth rate, this discovery having been made when someone thought that the residue from streptomycin manufacture might be a useful source of protein. Braude and his colleagues had also recently shown in controlled trials that copper salts, added to a pig diet at concentrations well beyond the level needed to satisfy physiological requirements, could also stimulate growth. It was argued that the response to antibiotic varied from station to station (what else would you expect?) and this might be due to differences in the resident bacterial flora at different stations. It was unknown whether the copper and the antibiotic responses were additive. Each response was of the order of 8%, averaged over a number of trials, which was close to the limit of detection. With a CV of 12% for growth rate, trials using 12–16 individually fed pigs would detect a true difference of 8% as significant at $P = 0.05$ in less than half of all trials conducted. Note that few trials would judge the effect of an antibiotic to be zero or negative, but a large proportion would find the apparent response 'not significant'. Also note that if 'unsuccessful' trials mostly go unreported, the evidence that gets published will give a seriously biased estimate of the true response.

Braude and his team decided to conduct a coordinated trial on as many dependable farms as could be recruited. The treatments were to be:

1. a control diet, suitable for growing pigs;
2. control + 250 mg Cu kg^{-1} (added as $CuSO_4$);
3. control + 10 mg oxytetracycline kg^{-1};
4. control + Cu + oxytetracycline.

Conditions for entry to the trial were that four pens of matched pigs could be provided (mostly 6–8 pigs in a pen), one pen for each treatment, and that feed intake and liveweight gain could be recorded weekly. Twenty-one centres agreed to join the trial. These were not drawn at random from pig farms up and down the country, as would have been theoretically desirable, but were all college or feed-company farms, since these had proved in earlier coordinated trials to be the only locations willing to cooperate and able to provide reliable records.

The growth response to copper (alone) in this experiment is illustrated in Fig. 14.1. The response to the antibiotic was smaller (although it had been 9.4% in a previous multi-location trial run by the same group) and the response to the combination of copper and antibiotic was the same as to copper alone.

The mean response to copper in this trial was a 9.7% increase in daily gain. However, the response estimated at individual centres ranged from −2% to +19%. You can imagine that the centre where the copper-treated group had grown more slowly than the controls was very sceptical about the benefits of added copper, while those who had recorded a 19% response were determined to use a copper supplement for all their growing pigs in future. This raises an important question as to whether the *local* result is more valuable than the *general* result for predicting future performance at a given location. Fortunately, 15 of the 21 centres were persuaded to repeat the trial in the following year, with very interesting results. The mean response to copper was again 9%, but there was no correlation between the results obtained at individual centres between one year and the next. One centre that recorded a negative response in the first year recorded an above-average result in the next year, and the centre with a 19% response in year 1 saw a slightly below average result in year 2. Of course, some centres recorded below-average responses in both years and a few were lucky enough to get above-average responses both times. That is the nature of random errors!

In the first year, with only one replicate of each treatment at each centre, the only error term available for testing treatment differences was the centre × treatment interaction. But, with 2 years' data, the error between pens at the same centre could be used to test the centre × treatment interaction. This interaction turned out to be zero. The variation that we see in Fig. 14.1 is simply the variation to be expected between replicated pairs of pens, one of which receives

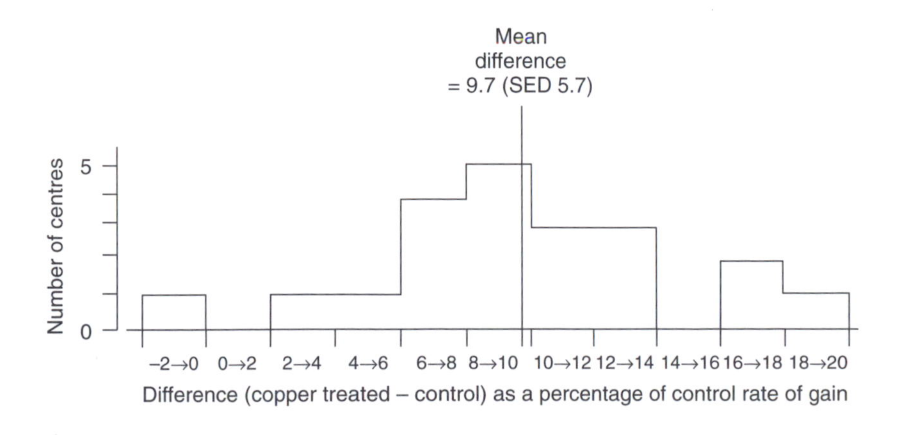

Fig. 14.1. The distribution of responses to copper sulphate added to a pig diet, as measured at 21 collaborating centres.

the control diet and the other the copper-supplemented diet. In this example, the conclusion is that the average result obtained from all centres is a better predictor of the future responses to be expected at one centre than the particular result obtain at that centre in 1 year.

The moral is not that treatment × location interaction is always absent, but that you should design your experiment so that you can test for it. This requires some replication at some locations (not necessarily all of them) so that a within-location error can be computed as well as the error that represents treatment variation from location to location. In an earlier pig trial by the same group of collaborators (Braude *et al.*, 1958) the level of feeding was varied and this resulted in marked differences in the response at different centres, not because of random errors but because of the differing growth potentials of the genotypes used.

Some other examples of trials that have been coordinated over multiple locations include the testing of mastitis control treatments in dairy cows (using the whole farm as the experimental unit, but switching treatments around between 1 year and the next) and testing the response of laying hens to an antibiotic added to the diet. In this latter case the experimental unit was a battery house filled with 15,000–30,000 laying pullets. Sites were chosen on which two similar houses could be allocated, one to treatment and one to control.

There are two possible reasons for going to the trouble and expense of coordinating a trial over many locations:

1. To obtain numbers of replicates on a scale not available at a single location.
2. To test whether an effect already recorded at one location can be reproduced under the varying conditions of a wider field to which the results may one day be applied.

These objectives are not mutually exclusive and an experimenter may choose a multi-location trial for both of these reasons. The final testing of new medicinal products is usually carried out on a wide population for both reasons. However, we can add a third reason for conducting large-scale medical trials: to discover whether there are occasional undesirable side-effects which would not be revealed by a small trial.

Should all trials involve multiple locations? Well no. Although testing an idea or a product at many locations adds considerably to your confidence that you have the right answer, the process is expensive and requires *more replication* for a given level of precision, because the variation amongst replicates is almost sure to be greater when replicates are widely scattered and involve differing genotypes, differing batches of feed, differing disease environments, differing managerial skills, etc.

There is a trade-off here. A multi-location trial gives *more generality* but *less precision* (for a stated number of replicates) than a trial conducted in one place at one time. For graduate students trying to gather enough results in a couple of years to make a thesis, precision is paramount. Everything possible must be done to exclude extraneous sources of variation. Establishing the generality of a result can then be left to an industrial sponsor who, you hope, will take up the results of your research. There is, however, a good case for saying that teams of established research workers should be more concerned than they sometimes are to conduct multi-location trials to see whether some interesting conclusion, which they have drawn from a single experiment, can be substantiated over a wider field. As noted at the beginning of this chapter, this usually happens by default in animal research work, because other competing scientists set out to 'repeat' unusual and interesting results after the initial paper has been published.

Reviewing Multiple Experiments

In the pig example described above, the *same treatments* were tested at a number of different centres and in two consecutive years. Such experiments can be conceived in advance and can be designed deliberately to recover information about treatment × location and treatment × year interactions. Another set of issues arises when faced with a series of experiments that attack the same question, whether the trials are your own or are culled from the literature, or are a combination of the two. Such experiments will often involve slightly different chosen treatments as well as utilizing different experimental material.

Analysis of a series of trials that bear upon the same question is called *meta-analysis*. This methodology has been particularly useful in medical research but has come into some disrepute in sociology and psychology because of lack of rigour in defining the question to be answered. If, for example, you take as your question 'Does psychoanalysis work?' and then try to summarize quantitatively all the trial reports involving psychoanalysis, you are wasting your time. It is easy to see that psychoanalysis may be helpful for some conditions but not for others and that whether or not it works must depend upon the criteria selected for judging the outcome. On the other hand, if we ask 'What is the relationship between metabolizable energy intake and milk energy output in multiparous non-pregnant dairy cows weighing about 500 kg?', we have a reasonable expectation that there is a graph or an equation that represents a general answer to this question. Similarly, a medical team can expect an answer to the question 'What is the benefit, as measured by increased 5-year survival rates, of a given dose of a given anti-cancer drug administered following surgical treatment to North American patients diagnosed with malignant melanoma?' Before rushing into meta-analysis, think about the nature of the question. Is it a single, tightly defined question or is it a rag-bag of questions? Is it reasonable to suppose that you can combine the results of experiments conducted at different places and at different times using different experimental material without taking account of those differences? Does the information available allow you to put the data into categories for analysis (e.g. experiments with Holsteins; experiments with Jerseys) and so test whether the different categories yield different outcomes?

Given that you are careful with your definition of the question and selective in your inclusion of material, analysis of a series of independent experiments can provide powerful answers which were not available before you made your analysis.

Selection of material must be done by using objective criteria which are set down for all to read. It will not do to pick out three experiments that support your thesis and to ignore four others that provide contrary evidence. One useful sieving criterion is to reject any report with less than a stated number of replications or with an SEM greater than a stated value. This gets rid of results that happen to have gone 'the wrong way' simply due to lack of precision. You will find an example of this approach in Morris (1968), where experiments testing the responses of laying hens to dietary energy concentration were included in the review only if they used at least 100 birds per dietary treatment.

Having assembled your material, the first step is to draw a graph of the data. This, if you are lucky, may suggest an emerging hypothesis. Two examples of raw data (though subjected to some trimming as discussed above) are shown in Figs 14.2 and 14.3.

The responses to temperature depicted in Fig. 14.2 were described as linear in almost all the original reports surveyed. What becomes clear from inspecting the whole body of evidence is that a curvilinear hypothesis is much more reasonable.

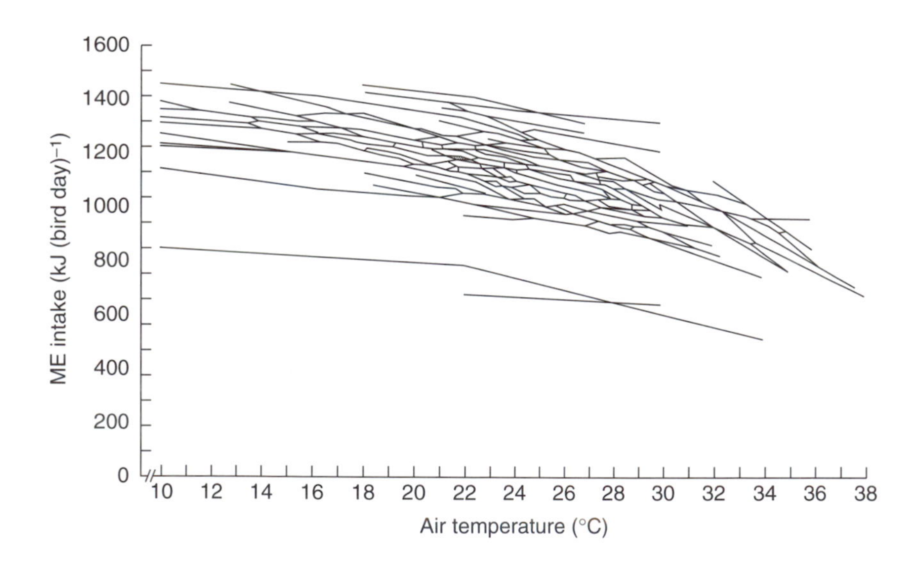

Fig. 14.2. Data from Marsden and Morris (1987) showing results of 61 previously published trials reporting the relationship between temperature and metabolizable energy (ME) intake in laying hens measured over a period of at least 8 weeks. Note that the response in nearly all trials is apparently linear, whereas the overall trend suggests an accelerating decline in feed intake as ambient temperature approaches body temperature.

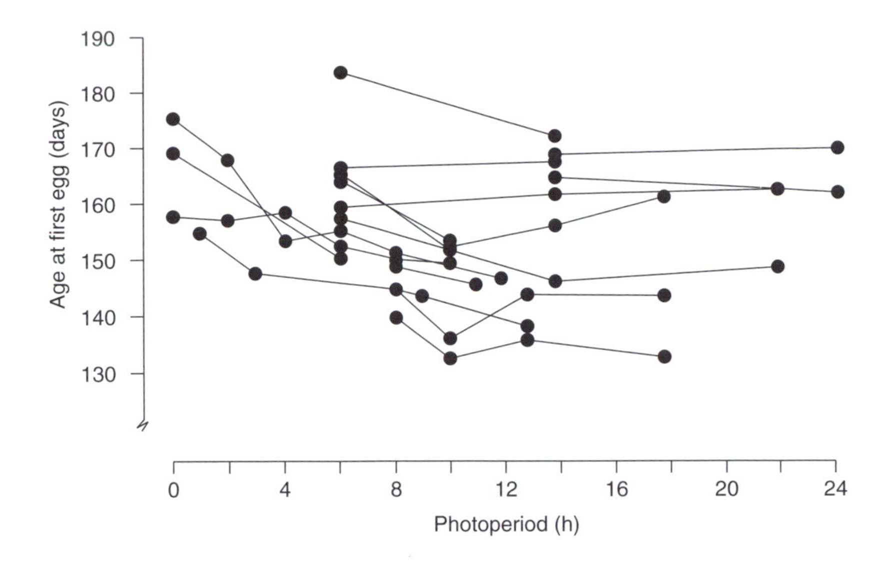

Fig. 14.3. Data reported by Lewis *et al.* (1998), representing all the available evidence showing the influence of photoperiod (held constant from hatching until after sexual maturity) on age at first egg in the fowl.

Now, you cannot simply average out the results reported at each temperature to derive a general trend because each experiment involves a different range of treatments and because there are differences in *level* of feed intake between trials. These differences in level may have simple explanations, such as the body size of the hens, their varying productivity or the nature of the diet; but you will not usually have the data or the necessary theory to allow you to adjust for each of these contributory factors. Figure 14.4 is drawn to illustrate just how far you can go wrong by simply taking averages. Because there is a marked difference in the level of appetite between the two trials depicted, simply averaging out the data produces a nonsense.

In the particular case illustrated in Fig. 14.4 you could resolve the problem by adjusting the appetite scale to read 'expressed as a percentage of voluntary feed intake when dietary ME = 9 MJ kg^{-1}'. But that will not work for Fig. 14.2 or 14.3 for two reasons. First, not all the trials have overlapping treatments. Secondly, each experimental point is subject to an error of estimate, and picking out one point on the abscissa as a fixed value to which all others are referred runs the risk that the selected value may have given rather atypical results in a particular experiment.

What we need to do is to lift or lower each line until they all fit together to give the most sensible and coherent picture, using *all the data* from a given trial to make the adjustment for that trial. To do this we first need to specify some model that we think represents the underlying response. In the case of the data in Fig. 14.2 this might be a parabola of the form:

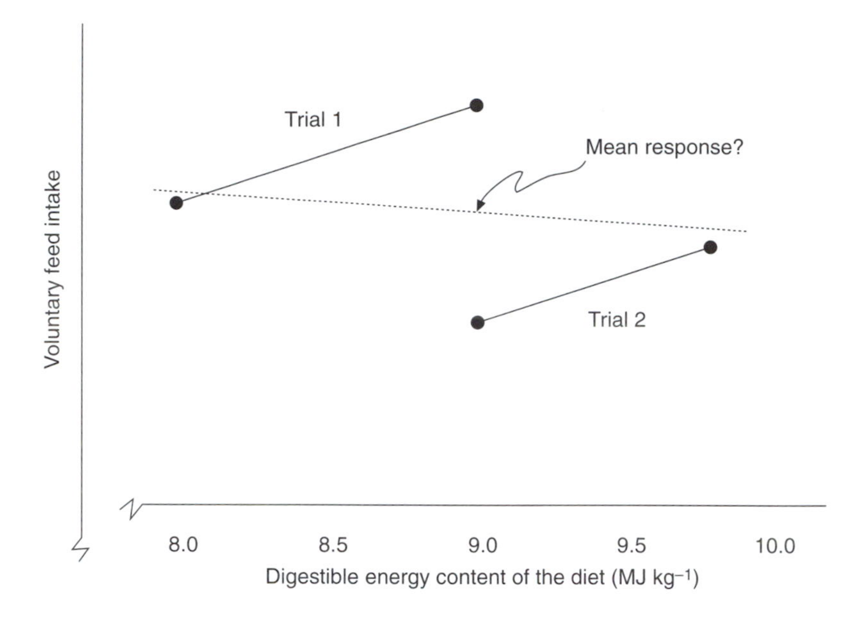

Fig. 14.4. A summary of two hypothetical experiments testing the effect of digestible energy content of ruminant diets on voluntary feed intake. In both trials the trend observed is positive; but calculating the means and fitting an average regression gives a negative slope.

$$y = a + b_1x - b_2x^2$$

where y = ME intake, and x = air temperature, although it could be a more complex model if the data can justify it.

To this model we add a term for each trial which represents the mean difference between that trial with its range of treatments and the general response observed over the full range of data. These *constants* fitted to represent mean differences between trials are not simple arithmetic differences between means of the data sets for the several experiments. A glance at Fig. 14.4 should convince you that adjusting those lines by the difference between the trial means will not resolve the problem. (It would if the two trials had identical treatments, but not otherwise.) In the case of Fig. 14.4 we need to fit *parallel* lines to the data for the two trials so as to arrive at a best estimate of the true slope of the relationship between feed intake and dietary digestible energy over the range from 8.0 to 9.75 MJ kg^{-1}.

Returning to Fig. 14.2, the first model we shall test is

$$y = a - b_1x - b_2x^2 + k_i$$

where k_i is a constant fitted to represent the ith trial; $i = 1 \ldots n$ and n is the number of trials.

This model allows us to fit 61 parallel curves to the sets of data in Fig. 14.2, with the condition that the residual sum of squared deviations of the observed values from the fitted curves is minimized. The final mean curve derived from this study is illustrated in Fig. 14.5.

The data in Fig. 14.3 present a similar-looking problem and indeed it is possible to fit a parabolic model to these points. However, further exploration shows, first, that adding a cubic term significantly improves the goodness of fit (because the curve is not symmetrical about its minimum point) and, secondly, that an even better fit can be obtained by adopting a hinged (bent-stick) model, as illustrated in Fig. 14.6.

The technique for fitting constants to remove mean differences between trials is merely an extension of regression analysis. With *n* experiments to

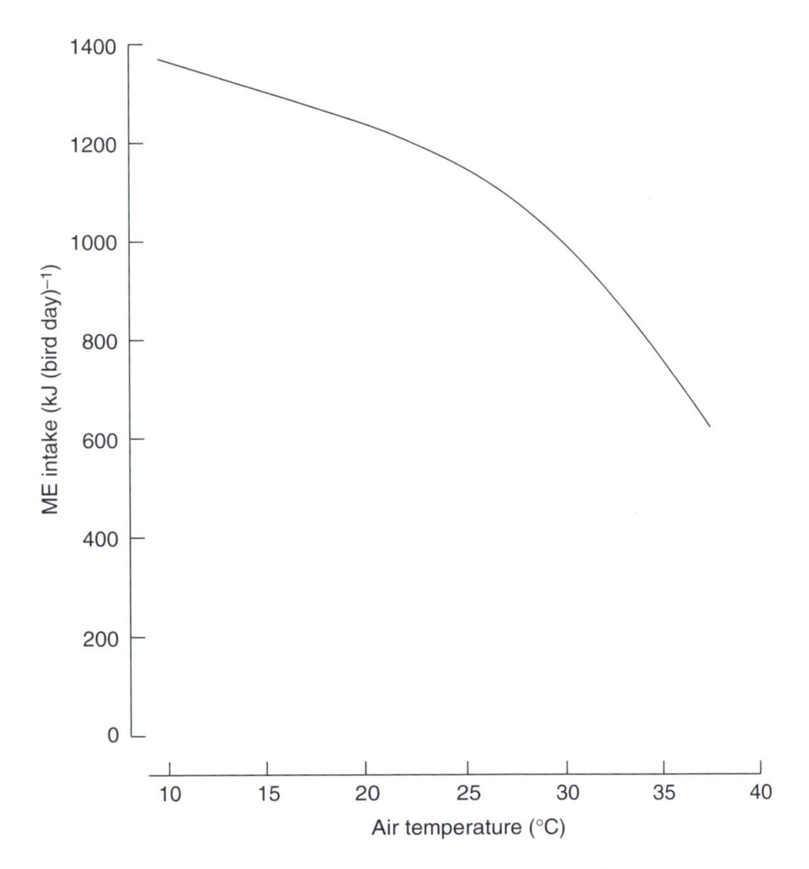

Fig. 14.5. A curve representing a condensation of the data illustrated in Fig. 14.2. The equation of the curve is:

$$y \text{ (ME intake, kJ day}^{-1}) = 1606 - 35.28T + 1.647T^2 - 0.0362T^3$$

where T = air temperature. The cubic equation is a significantly better fit than a quadratic in this case.

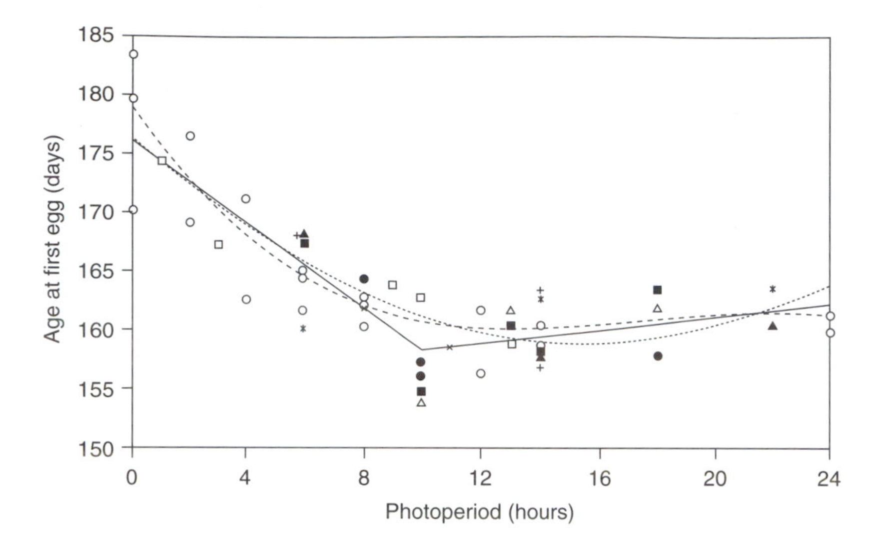

Fig. 14.6. Three models – a quadratic curve, a cubic curve and a bent-stick – fitted to the data represented in Fig. 14.3. In this case the bent-stick model turned out to be a better fit than either of the smooth curves.

summarize you will need $n - 1$ dummy variables to represent the difference between the first trial in your list and each of the other trials. The fitting of a general mean effectively sets a level for the first trial and the constants then represent the appropriate adjustments to that level for each succeeding trial. The matrix of numbers required to analyse the data in Fig. 14.3 is set out in Appendix 23 as an example of the methodology.

Meta-analysis is a powerful tool and should be used much more extensively by graduate students as a means of reviewing the literature than it currently is. Do try to replace pages and pages of paragraphs summarizing what others have reported and concluded with a well-judged series of graphs that represent all the previously published evidence that is worthy of consideration. Such graphs will often lead to models in which mean differences between trials are removed by fitting constants, thus giving a succinct summary in one equation of what may at first seem a very diverse set of responses when described in words. Note, for example, that if Fig. 14.6 represents a true relationship, it is very easy for three experimenters, each comparing two treatments, to conclude that sexual maturity in chickens is negatively related to daylength, or positively related, or that photoperiod has no influence at all on age at first egg.

As with all powerful tools, there are dangers. The chief danger lies in not having thought carefully enough about the rigorous definition of your question. For example, you may be interested in the response (measured as heat output) of animals to variation in ambient temperature: but there are two quite different responses here. In the short term, homeotherms respond to heat stress by panting, which increases their heat output; but in the longer term they respond to heat stress by reducing their feed intake, their activity and their output (in the form of growth, milk yield or egg yield, as the case may be), with the result that heat output is lower under prolonged heat stress than it is for similar animals at a comfortable temperature. To review this particular field it is essential to draw the distinction between short-term and long-term (adapted) responses.

Thus you need to examine the results of other people's experiments carefully to see what the data are telling you, before lumping all their experiments together in a meta-analysis. Where the data can be arranged in different categories, such as males and females, mature cows and heifers, inbred mice and non-inbred mice, first examine the results in the separate categories to see whether the responses are similar. If you have enough material, undertake the statistical tests required to see whether the responses are homogeneous across categories. Slopes of responses may vary for good reasons and, if you are persistent and a bit lucky, you may be able to discover those reasons. For example, the relationship between digestible energy content of the diet and feed intake portrayed in Fig. 14.4 is only indirectly a causal one. Diets of lower digestibility take longer to clear the rumen than highly digestible ones and that is the reason for the regression observed; but the slope is not a constant fixed by some physical law and may well vary with the precise nature of the diets that are being tested. Diets (or feeding strategies) therefore need to be assigned to categories before looking for a rate of response that can be useful as a predictor in future cases.

Summary

1. Multi-location trials are *costly* and tedious to organize, *but* are capable of producing *powerful* general answers to important questions.
2. There is an inevitable *antagonism* between *precision* and *generality* when choosing an experimental protocol. As more genotypes, more varied diets, more locations and more seasons are included in the design, so the experiment becomes more variable (demanding greater replication), but also much more useful.
3. The analysis of previously published trials should always be done quantitatively, with a view to producing a *new synthesis*.
4. Combination of results by *fitting constants* to remove mean differences between trials is a powerful, and underused, technique.

References

Braude, R., Townsend, M.J., Harrington, G. and Rowell, J.G. (1958) A large scale test of the effects of food restriction on the performance of fattening pigs. *Journal of Agricultural Science (Cambridge)* **51,** 208–217.

Braude, R., Townsend, M.J., Harrington, G. and Rowell, J.G. (1962) Effects of oxytetracycline and copper sulphate, separately and together, in the rations of growing pigs. *Journal of Agricultural Science (Cambridge)* **58,** 251–256.

Lewis, P.D., Morris T.R. and Perry, G.C. (1998) A model for the effect of constant photoperiods on the rate of sexual maturation in pullets. *British Poultry Science* **39,** 147–151.

Marsden, A. and Morris, T.R. (1987) Quantitative review of the effects of environmental temperature on food intake, egg output and energy balance in laying pullets. *British Poultry Science* **28,** 693–704.

Morris, T.R. (1968) The effect of dietary energy level on the voluntary calorie intake of laying birds. *British Poultry Science* **9,** 285–295.

Epilogue

If you have read all of the book thus far, you have done well. There are 26 appendices to follow, but one hopes that you have already dipped into those while reading the preceding chapters.

If there is a general message in this book, it is that you should take every opportunity to explore statistical techniques for analysing your data and, if you are lucky enough to have access to a statistician, you should certainly make full use of professional statistical advice wherever possible. Having done all that, you must then strive to make up your own mind about the commonsense, logical conclusions that can be drawn from your data. Do not be hidebound by statistical tests, and remember that statistics can only assign probabilities to your hypotheses, never certainties. Indeed the exciting thing about science is that nothing is ever certain – it is simply the best explanation of natural observations that we can come up with for the time being. The world got along very well for 250 years by assuming that Newton's laws of motion were 'true' and, even when Einstein showed that he was wrong in some respects, Newtonian mechanics were still good enough to land a man on the moon and to bring him home again.

Appendix 1

Random Numbers and How to Use Them

The following table of random numbers is taken from Mead, Curnow and Hasted's *Statistical Methods in Agriculture and Experimental Biology*, published by Chapman & Hall, London, with kind permission from Kluwer Academic Publishers.

10 27	53 96	23 71	50 54	36 23	54 51	50 14	28 02	12 29	88 87
85 90	22 58	52 90	22 76	95 70	02 84	74 69	06 13	98 86	06 50
44 33	29 88	90 49	07 55	69 50	20 27	59 51	97 53	57 04	22 26
47 57	22 52	75 74	53 11	76 11	21 16	12 44	31 89	16 91	47 75
03 20	54 20	70 56	77 59	95 60	19 75	29 94	11 23	59 39	14 47
64 17	18 43	97 37	66 55	86 08	74 50	43 43	23 29	16 24	15 62
91 14	61 71	03 40	15 69	44 46	54 66	35 01	87 61	23 76	36 80
27 71	29 93	52 89	64 78	32 97	65 28	99 82	41 10	97 52	41 91
12 96	17 70	72 76	17 93	38 26	72 96	28 73	27 64	78 16	72 81
54 30	61 13	60 50	61 56	40 20	19 22	30 61	43 89	60 09	82 39
83 32	99 29	30 06	19 71	11 32	69 17	86 34	50 76	37 41	76 54
27 17	25 61	91 76	19 54	99 73	97 21	44 87	39 63	24 22	74 30
40 89	21 88	56 84	11 75	74 88	23 55	48 98	19 48	79 81	92 62
51 66	17 48	29 96	00 83	81 23	58 09	21 39	39 20	83 46	30 75
95 22	63 34	58 91	78 22	50 22	77 21	14 19	58 66	49 25	03 51
93 83	73 70	80 88	71 85	64 44	57 50	19 82	60 77	38 95	93 33
42 02	33 18	33 55	96 66	88 38	16 80	77 51	17 96	49 76	99 28
42 42	13 33	66 00	18 37	58 80	54 32	00 96	25 16	15 37	34 12
66 71	67 54	79 25	64 34	82 15	28 97	88 84	84 51	62 90	17 71
73 05	53 85	63 18	06 47	71 00	32 31	59 72	34 28	70 83	12 90
02 80	12 24	34 78	22 50	57 02	07 01	13 00	78 80	94 93	14 53
22 89	81 32	32 72	48 92	95 75	88 56	75 53	79 17	53 81	54 17
95 45	64 84	17 28	06 57	71 96	81 36	37 65	42 62	43 84	45 23
10 30	05 07	21 34	59 18	85 95	21 87	73 16	78 37	15 98	16 66
73 39	21 94	01 84	28 20	50 35	57 82	88 13	52 53	76 73	68 22
47 91	87 36	45 69	03 01	24 25	13 64	42 74	36 67	77 67	00 92
39 24	26 77	62 37	82 46	93 96	82 75	75 16	95 05	30 68	83 02
77 29	09 12	41 77	29 57	34 89	94 95	45 70	59 85	38 04	04 80
04 78	20 07	17 15	68 12	38 26	01 90	68 30	83 80	19 89	98 65
83 81	53 08	09 23	22 61	99 41	27 90	35 43	07 09	62 26	45 83
97 67	74 54	96 14	63 28	98 11	18 33	82 60	90 41	33 11	77 59
52 80	26 89	13 38	70 08	73 22	64 70	83 44	49 24	20 93	12 59
80 69	43 27	33 56	39 88	73 31	24 44	87 33	08 21	40 06	77 91
00 48	24 08	73 92	37 19	69 87	91 79	86 27	47 91	31 70	53 52
14 91	97 37	53 40	46 26	29 25	96 42	57 22	94 34	59 71	23 59
50 62	28 51	94 10	15 18	06 02	39 94	13 91	54 50	60 27	26 68
17 59	53 08	58 06	80 00	75 71	95 13	76 91	24 55	34 09	97 12
73 17	99 45	85 28	63 17	99 31	24 62	75 82	78 89	27 59	18 62
37 95	74 96	25 44	95 66	42 02	31 48	82 21	76 87	86 75	07 95
76 95	18 76	76 28	18 60	44 92	76 09	46 96	39 37	27 12	30 44

Example: To allocate four treatments at random within three blocks

1. Select a position on the page at random (shut your eyes and stab with a pin): suppose your pin marks a point at the top of the section boxed in.
2. Read off pairs of numbers down the column from your randomly selected starting point (continue at top of next column if necessary).
3. Divide these two-digit numbers by the number of treatments (four in this example) and list the remainders. Use these remainders (which will run from 0 to 3) to give you the sequence of treatments within each block as illustrated below:

Remainder after dividing random number by 4	Treatments allocated in order
1	**1**
0	**4**
2	**2**; and the last treatment in block 1 must be **3**
0	**4** (first treatment in block 2)
0	(4 again – ignore this)
3	**3**
0	(4 again – ignore this)
1	**1**; and the last treatment in block 2 must be **2**
1	**1**
1	(1 again – ignore this)
1	(1 again – ignore this)
2	**2**
2	(2 again – ignore this)
1	(1 again – ignore this)
3	**3**; and the last treatment in block 3 must be **4**

If you have three treatments to allocate, you should use numbers 01–99 and ignore 00. If you have six treatments, use 01–96 and ignore numbers 97–00 when they turn up.

Appendix 2

Some Useful General Formulae

The standard error of a mean (which in this book is labelled SEM, but is also commonly abbreviated as SE or s.e.) is calculated as

$$\text{SEM} = \sqrt{(s^2/n)} = s/\sqrt{n}$$

where s = the standard deviation (of replicate values about the mean); and n = the sample size.

If you have performed an ANOVA (which assumes that all treatments have the same variance), then the s^2 you are looking for is the error mean square and, in calculating SEM for a treatment mean, n is the number of replicates contributing to that mean.

The standard error of the difference between two means (SED in this book but also sometimes confusingly given as s.e. in published tables and computer printouts) is calculated as the square root of the sum of the squares of the standard errors of the two means, that is:

$$\begin{aligned}\text{SED} &= \sqrt{[(\text{SEM}_1)^2 + (\text{SEM}_2)^2]} \\ &= \sqrt{[(s_1^2/n_1) + (s_2^2/n_2)]}.\end{aligned}$$

The above formula must be used if the two treatments being compared have unequal variances. However, if you have performed an ANOVA and obtained a pooled error m.s. which is presumed to be applicable to all treatments, and *if* the treatments all have the same number of replicates, then $s_1^2 = s_2^2$ and $n_1 = n_2$ and therefore

$$\text{SED} = \sqrt{(2s^2/n)}.$$

Also note that

$$\text{SED} = \sqrt{2}\,.\,\sqrt{(s^2/n)} = \sqrt{2}.(\text{SEM}).$$

Useful Quick Approximations

If a published table gives an SED then the approximate least significant difference (LSD) between two means (at $P = 0.05$) can be estimated quickly by assuming that $t = 2$, which is correct for 60 d.f. and near enough for practical purposes for fewer d.f. Thus:

$$\text{LSD} \approx 2\ (\text{SED}).$$

If a published table gives an SEM, common to all means in the row or column, then the approximate LSD between any two means (at $P = 0.05$) can be quickly calculated as:

$$\text{LSD} = \sqrt{2}.t.\text{SEM}$$

$$\approx 3\ (\text{SEM}).$$

Differences between Differences

The principles given above apply to finding the appropriate SED for a difference between two differences.

Given four means, A, B, C and D each with their own SEM,

SED_{A-B}, for the difference A − B = $\sqrt{(\text{SEM}_A^2 + \text{SEM}_B^2)}$

SED_{C-D}, for the difference C − D = $\sqrt{(\text{SEM}_C^2 + \text{SEM}_D^2)}$.

If you then wish to ask whether the difference (A − B) differs significantly from the difference (C − D), the appropriate SED is:

$$\sqrt{(\text{SED}_{A-B}^2 + \text{SED}_{C-D}^2)} = \sqrt{(\text{SEM}_A^2 + \text{SEM}_B^2 + \text{SEM}_C^2 + \text{SEM}_D^2)}.$$

If, in a particular case, the four means A, B, C and D all have the same estimated variance, s^2, and are all based on the same number of replications, n, then the SED for the difference between two differences is:

$$\text{SED}_{(A-B)-(C-D)} = \sqrt{(4s^2/n)}.$$

Appendix 3

Answers to Exercise 2.1 Using a Pocket Calculator

The carcass weights (kg) are repeated here (Table A3.1) for convenience.

Table A3.1. Data from Exercise 2.1.

Block No.	Basal diet	Basal + acetate	Basal + propionate	Basal + butyrate	Block totals
I	16.3	18.9	19.4	18.0	72.6
II	16.4	18.2	17.6	17.5	69.7
III	16.7	18.9	17.6	18.6	71.8
IV	17.7	19.5	19.8	19.1	76.1
V	18.0	17.4	19.3	18.4	73.1
VI	19.1	18.0	16.5	17.6	71.2
VII	19.1	21.0	18.9	21.3	80.3
VIII	18.0	21.3	19.9	21.1	80.3
Diet totals	141.3	153.2	149.0	151.6	595.1 (Grand total)

1. Calculate the correction factor (CF)

$$= \frac{(\text{Grand total})^2}{N} = \frac{(595.1)^2}{32} = 11{,}067.0003$$

(N = total number of plots in the experiment).

2. Calculate the blocks sum of squares (s.s.)

$$= \frac{\Sigma[(\text{block totals})^2]}{n_t} - \text{CF} = \frac{44384.53}{4} - 11067.0003 = \mathbf{29.1322}.$$

(n_t = number of treatments = number of plots in each block total).

3. Calculate the Diets s.s.

$$= \frac{\Sigma[(\text{diet totals})^2]}{n_b} - \text{CF} = \frac{88619.49}{8} - 11067.0003 = \mathbf{10.4359}$$

(n_b = number of blocks = number of plots in each diet total).

4. Calculate the total s.s.

$$= \Sigma[(\text{plot values})^2] - \text{CF} = 11125.93 - 11067.0003 = \mathbf{58.9297}.$$

5. Calculate the error s.s.

$$= \text{total s.s.} - (\text{blocks s.s.} + \text{diets s.s.})$$

$$= 58.9297 - (29.1322 + 10.4359) = \mathbf{19.3616}.$$

6. Enter the s.s. in an ANOVA table and calculate mean squares; m.s. = s.s./d.f.

7. Calculate F values as the ratio of each m.s. to the error m.s.

8. Check the F-ratios against the values tabulated in Appendix 25 to obtain the corresponding probability estimates.

Table A3.2. The completed ANOVA.

Source	d.f.	s.s.	m.s.	F-ratio
Blocks	7	29.1322	4.1617	4.51**
Diets	3	10.4359	3.4786	3.77*
Error	21	19.3616	0.92198	
Total	31	58.9297		

* = $P < 0.05$.
** = $P < 0.01$.

Appendix 4

Answers to Exercise 2.1 Using SAS[a]

For more detailed instructions and help with the use of SAS, you should consult specific SAS manuals, tutorials and help files.

The first statement in Table A4.1 creates a data file named 'AE22'. Note the use of a semicolon (;) at the end of every SAS execution or procedure statement. The succeeding input statement identifies the names of variables and their location in your data set. The '$' sign tells SAS that the diet variable is not numeric, but descriptive. The executable statements which follow the input statement create a numeric variable now called 'TREAT'. Executable statements can also be used to create new variables (e.g. log x) as part of the input step and additional variables can be created by, for example, merging two previously loaded sets of data.

Your data follow after a 'CARDS;' statement. This nomenclature is a throwback to the days when data were entered by means of punched cards. Data can also be imported from external files, such as Excel files (more of this later). SAS reads the data in the order encountered, with no indication of data location. However, you could assign column numbers for each variable by entering:

```
INPUT SEX 1 BLOCK 3 DIET $ 5–6 CWT 8–11;
```

This is helpful if you have missing observations. If you enter missing data as blank spaces and use normal listing, SAS will overlook the space and subsequent data entry will then be incorrect. The best procedure is always to enter a full stop to mark any missing observation, as this will work whichever method of data entry you use. Using a 'RUN;' statement completes the entry step, although it is not essential since the task will be executed when the next procedure statement is encountered. However, two procedure programs are then loaded into the memory simultaneously and, for older PCs, using the 'RUN;' statement after each procedure or input step is more efficient.

Printing the data and checking that it has been correctly entered is always the first step when running a SAS job. To do this you enter:

[a] SAS has been chosen to illustrate the use of a computer package to solve some of the exercises in this book, not because the author regards it as the best program for statistical analysis, but because it is one of the most widely used programs on the world scene.

Table A4.1. The entry of data from Exercise 3.1 for analysis by SAS.

```
DATA AE22;
INPUT SEX BLOCK DIET $ CWT;
IF DIET='CON' THEN TREAT=1;
IF DIET='ACE' THEN TREAT=2;
IF DIET='PRO' THEN TREAT=3;
IF DIET='BUT' THEN TREAT=4;
CARDS;
1 1 CON 16.3
1 2 CON 16.4
1 3 CON 16.7
1 4 CON 17.7
2 5 CON 18.0
2 6 CON 19.1
2 7 CON 19.1
2 8 CON 18.0
1 1 ACE 18.9
1 2 ACE 18.2
1 3 ACE 18.9
1 4 ACE 19.5
2 5 ACE 17.4
2 6 ACE 18.0
2 7 ACE 21.0
2 8 ACE 21.3
1 1 PRO 19.4
1 2 PRO 17.6
1 3 PRO 17.6
1 4 PRO 19.8
2 5 PRO 19.3
2 6 PRO 16.5
2 7 PRO 18.9
2 8 PRO 19.9
1 1 BUT 18.0
1 2 BUT 17.5
1 3 BUT 18.6
1 4 BUT 19.1
2 5 BUT 18.4
2 6 BUT 17.6
2 7 BUT 21.3
2 8 BUT 21.1
RUN;
```

```
PROC PRINT; RUN;
```

You may also want to use 'PROC PLOT;' to allow you to make a visual appraisal of your data before proceeding further. Are there outliers? What patterns of response are evident (or absent)? How much variation do you see?

Are there mistakes in inputting, analysing or transcribing your observations? Remember that the scale of the axes affects your impression of the data. If plotting more than one group of data, you may want to use the '`PROC PLOT UNIFORM;`' statement to ensure that all graphs with the same variables use the same scales.

The instruction:

```
PROC MEANS; RUN;
```

will give you an overall mean with default parameters for each variable, but you can also specify which parameters you want calculated, for example:

```
PROC MEANS MEAN N STDERR STD T PRT; RUN;
```

Before generating means for individual class variables the data must be sorted. This is done by:

```
PROC SORT; BY DIET; RUN;
```

Note that the diet means will now be sorted in alphabetical, not numerical order. This can be important if you are importing means generated by SAS directly into a results table using Word or another Windows-based package and you want the treatments listed in some order other than alphabetical.

To obtain parameters for each diet, enter:

```
PROC MEANS MEAN N STDERR STD T PRT; BY DIET; RUN;
```

Analysis of variance can be carried out using a number of programs in SAS, but '`PROC GLM;`' (GLM stands for general linear model) has more options and is always used if the data set has missing observations, because it generates least squares means (see Chapter 11). Enter:

```
PROC GLM;
CLASSES BLOCK SEX TREAT;
MODEL CWT=BLOCK TREAT;
LSMEANS TREAT/STDERR;
QUIT;
```

This model will provide the solution to Exercise 2.1. The '`CLASSES;`' statement identifies the discrete variables used in this ANOVA. An '`LSMEANS;`' statement asks that the least squares means be generated, and adding the '`/STDERR;`' option asks for the appropriate SEM (from the ANOVA) to be printed. The SEM will, of course, be the same for each treatment if they have equal numbers of observations.

The completed ANOVA is given in Appendix 3, Table A3.2.

Appendix 5

Answers to Exercise 2.2 Using a Pocket Calculator

The data from Exercise 2.2 are repeated in Table A5.1.

Table A5.1. Data from Exercise 2.2. Dressed carcass weight (kg).

	Block no.	Basal diet	Basal + acetate	Basal + propionate	Basal + butyrate	Block totals
Ewes	I	16.3	18.9	19.4	18.0	72.6
	II	16.4	18.2	17.6	17.5	69.7
	III	16.7	18.9	17.6	18.6	71.8
	IV	17.7	19.5	19.8	19.1	76.1
Ewe subtotals		(67.1)	(75.5)	(74.4)	(73.2)	(290.2)
Wethers	V	18.0	17.4	19.3	18.4	73.1
	VI	19.1	18.0	16.5	17.6	71.2
	VII	19.1	21.0	18.9	21.3	80.3
	VIII	18.0	21.3	19.9	21.1	80.3
Wether subtotals		(74.2)	(77.7)	(74.6)	(78.4)	(304.9)
Treatment totals		141.3	153.2	149.0	151.6	595.1

1. Set out a framework for the ANOVA as in Table A5.2 so that you are clear what variance components you need to calculate.

Table A5.2. Framework for the ANOVA.

Source	d.f.	
Blocks	7	
Sex		1
Blocks within sexes		6
Diets	3	
Blocks × diets	21	
Sex × diet		3
Residual = error = diets × blocks within sexes		18
Total	31	

Note that the sex s.s. is *part* of the blocks s.s. and not additional to it (it is, in fact, a comparison of blocks I–IV *versus* blocks V–VIII). The error term of the simple randomized block design has been re-labelled 'blocks × diets' and this component is divided into a sex × diet interaction and the remainder, which is the new error term.

2. Sums of squares for blocks, diets, blocks × diets (labelled 'Error' in Exercise 2.1) and total are the same as before and can be copied from Appendix 3.

3. Calculate sex s.s. = $\frac{\Sigma[(\text{sex totals})^2]}{n} - \text{CF}$

(**N.B.** n = number of plots in each sex total = 16, *not* 2 = number of sexes)

$$= \frac{290.2^2 + 304.9^2}{16} - 11067.0003 = \mathbf{6.7528}.$$

4. Calculate blocks within sexes s.s. either

(a) by difference: blocks within sexes s.s. = blocks s.s. − sex s.s.

$= 29.1322 - 6.7528 = \mathbf{22.3794}$, or

(b) in two steps:

(i) blocks within ewes s.s.

$$= \frac{\Sigma[(\text{ewe block totals})^2]}{n = \text{no. in each total}} - \frac{(\text{ewe total})^2}{n = \text{no. in total}}$$

$$= \frac{72.6^2 + \ldots 76.1^2}{4} - \frac{290.2^2}{16} = \mathbf{5.3225}.$$

(ii) blocks within wethers s.s., similarly

$$= \frac{73.1^2 + \ldots 80.3^2}{4} - \frac{304.9^2}{16} = \mathbf{17.0569}.$$

Blocks within sexes s.s. = 5.3225 + 17.0569 = **22.3794**.

5. Calculate the sex × diet interaction s.s. As with any interaction, this is a two-step calculation. First calculate the s.s. for the cells in a two-way table (Table A5.3) representing *totals* for the two interacting factors.

Table A5.3. The two-way table needed for calculating interaction.

	Diet				
Sex	Bas.	Ac.	Pr.	Bu.	Sex totals
Ewes	67.1	75.5	74.4	73.2	290.2
Wethers	74.2	77.7	74.6	78.4	304.9
Diet totals	141.3	153.2	149.0	151.6	595.1

The variation in this table, which we shall call the total treatment s.s., is

$$\frac{67.1^2 + 75.5^2 + \ldots 74.6^2 + 78.4^2}{4\ (= \text{no. of plots in each total})} - \text{CF} = 20.7272.$$

This term represents variation due to sexes and diets *and* the interaction.

Sex × diet interaction s.s. (3 d.f.)

= total treatment s.s. (7 d.f.) − {sex s.s. (1 d.f.) + diet s.s. (3 d.f.)}

= 20.7272 − (6.7528 + 10.4359) = **3.5385**.

6. The residual error s.s. (which can also be labelled blocks within sexes × diet s.s.) has 6 × 3 = 18 d.f. and is found by difference.

Blocks within sexes × diet s.s. = blocks s.s. − sex × diet s.s.

= 19.3616 − 3.5384 = **15.8232**.

7. Enter the s.s. in an ANOVA table, remembering to offset the entries where a component has been split into subcomponents so that the d.f. and s.s. add to the correct totals (Table A5.4).

Table A5.4. The completed ANOVA.

Source	d.f.		s.s.		m.s.	*F*-ratio
Blocks	7		29.1322			
Sex		1		6.7528	6.7528	7.68*
Blocks within sexes		6		22.3794	3.7299	4.24*
Diets	3		10.4359		3.4786	3.95*
Blocks × diet	21		19.3616			
Sex × diet		3		3.5384	1.1795	1.34
Residual (= error)		18		15.8232	0.87907	
Total	31		58.9297			

8. For a subdivision of the diet s.s in this example, see Chapter 3 and Appendix 7.

Appendix 6

Answers to Exercise 2.2 Using SAS

The method of entering data for SAS is given in Appendix 4. Having entered your data and checked it, the following program will give you the output to answer Exercise 2.2.

```
PROC GLM;
CLASSES BLOCK SEX TREAT;
MODEL CWT=SEX BLOCK(SEX) TREAT SEX*TREAT;
LSMEANS TREAT/STDERR;
QUIT;
```

Because this example contains no missing plots, you can also use 'PROC ANOVA' to obtain the same results.

```
PROC ANOVA;
CLASSES BLOCK SEX DIET;
MODEL CWT=SEX BLOCK(SEX) DIET SEX*DIET;
MEANS DIET SEX SEX*DIET;
QUIT;
```

In this case the variable 'DIET' has been used, which will give a different order of listing for the means in your output.

The completed ANOVA is given in Appendix 5, Table A5.4.

Appendix 7

Separation of Treatment Means

In Examples 2.1 and 2.2 there were four treatments, a basal diet and three diets with supplementary volatile fatty acid (VFA) salts. Suppose that we wish to make the comparisons:

1. Basal versus the three VFA supplemented diets.
2. Amongst the three VFA salts.

This can be done by the general methods for calculating s.s. given in Appendices 3 and 5.

$$\text{Basal versus rest s.s.} = \frac{(\text{basal total})^2}{n \text{ in total}} + \frac{(\text{total of VFA diets})^2}{n \text{ in total}} - \text{CF}$$

$$= \frac{(141.3)^2}{8} + \frac{(453.8)^2}{24} - \frac{(595.1)^2}{32} = 9.3126$$

$$\text{Amongst VFA s.s.} = \frac{\Sigma[(\text{VFA totals})^2]}{n \text{ in each total}} - \frac{(\text{total of VFA diets})^2}{n \text{ in total}}$$

$$= \frac{\Sigma[153.2^2 + \ldots 151.6^2]}{8} - \frac{(453.8)^2}{24} = 1.1233.$$

You will have noticed that the amongst VFA s.s. could have been found by subtraction from the diet s.s.:

basal versus rest s.s. + amongst VFA s.s. = diet s.s.

$$9.3126 \quad + \quad 1.1233 \quad = 10.4359$$

Another method of calculating the first of these s.s. is by means of *orthogonal polynomials*. This trick is worth learning, because it will allow you to:

1. construct a set of questions which it is statistically sound to ask of your data; and
2. quickly find the corresponding component s.s.

Orthogonal Polynomials

First we will illustrate, with a simple example, how these work and then we will give the general rules that apply to their use in separating treatment means.

Suppose that we wish to find the s.s. for Exercise 1.2 corresponding to the contrast:

basal diet *versus* mean of the other three diets.

In numerical terms (and using totals rather than means because that is more convenient) this effect is represented by

$$141.3 - \tfrac{1}{3}(153.2 + 149.0 + 151.6)$$

or, multiplying by 3 to get rid of fractions,

$$+3(141.3) - 1(153.2) - 1(149.0) - 1(151.6) = \mathbf{-29.9}.$$

We call this number the contrast (C) which we are looking for and the corresponding s.s. is

$$C^2/(\sum c^2 \times n)$$

where c represents the coefficients used (+3, −1, −1 and −1 in this case) and n = the number of plots in each total used (8 in this case).[a]

Since $3^2 + (-1)^2 + (-1)^2 + (-1)^2 = 12$, the s.s. we are looking for $= (-29.9)^2/(12 \times 8) = \mathbf{9.3126}$.

Although the explanation takes some time to set out, the calculation of the s.s. from the treatment totals is quicker than by the first method given in this appendix. Moreover, it becomes much quicker if we wish to evaluate other contrasts at the same time.

Suppose that we decide to look for three contrasts:

1. basal versus the rest;
2. acetate plus butyrate versus propionate;
3. acetate versus butyrate.

To find the corresponding s.s. we set out Table A7.1.

This is perhaps a good moment to point out that the *scale* of the coefficients does not matter (had we written $+1, -\frac{1}{3}, -\frac{1}{3}, -\frac{1}{3}$ in column **1** we should have obtained $\sum c^2 = 1\frac{1}{3}$, $C = -9.9667$ and the s.s. would still be 9.3126): neither does the *sign* of the coefficients matter (had we written −1, +2, −1 in column 2, $\sum c^2$ would still be 6 and C^2 would be $(-6.8)^2$ which is the same as 6.8^2). It is, however, essential that we put *contrasting* signs on the quantities that we wish to contrast and it is essential that the coefficients sum to zero.

[a] Because this method of calculating s.s. involves taking *differences* between treatment totals, there is no correction for the mean (CF) as in the method used at the beginning of this appendix, which is based on *adding* squares of treatment totals.

Table A7.1. Orthogonal polynomials for selected contrasts in Exercise 2.2.

	Treatment totals (n = 8)	Contrast 1	2	3
Basal	141.3	+3		
Acetate	153.2	−1	+1	+1
Propionate	149.0	−1	−2	
Butyrate	151.6	−1	+1	−1
	Σc =	0	0	0
	Σc^2 =	12	6	2
	C =	−29.9	6.8	1.6
	s.s. = $C^2/(\Sigma c^2 \,.\, 8)$ =	**9.3126**	**0.9633**	**0.1600**

Now we come to the rules for making up an orthogonal[a] set of contrasts, using coefficients (or polynomials) as above.

Rule 1. Any contrast must be between two quantities and thus represent 1 d.f. only.
Rule 2. The maximum number of contrasts available in one orthogonal set is equal to the d.f. available.
Rule 3. To be a valid contrast, the coefficients must sum to zero ($\Sigma c = 0$).
Rule 4. For one contrast to be orthogonal to another, the sum of the products of their coefficients must be zero $\{\Sigma(c_1.c_2) = 0\}$.

If we test this last rule on the table of coefficients (Table A7.1), we obtain for columns **1** and **2**:

$$\Sigma[c_1.c_2] = (+3)(0) + (-1)(+1) + (-1)(-2) + (-1)(+1) = 0$$

and therefore contrasts 1 and 2 are orthogonal.

Applying the same test to columns **2** and **3** we have:

$$\Sigma[c_2.c_3] = (+1)(+1) + (-2)(0) + (+1)(-1) = 0.$$

You will find, in any set, that if column A is orthogonal with B and B is orthogonal with C, then A and C will also be mutually orthogonal. It is thus only necessary to check that *one column* is orthogonal with *each of the others* to know that you have an *orthogonal set*.

If a set of contrasts obey the rules for orthogonality given above, then the corresponding component s.s. will add to the treatment s.s. You will have noticed that:

9.3126 + 0.9633 + 0.1600 = **10.4359** (the diets s.s.).

We can now illustrate, in Table A7.2, why testing all possible pairs from a four-treatment experiment does not provide an orthogonal set of contrasts.

[a] The word 'orthogonal' means at right angles or 'all square'. In this context, it means having the proper relationship to each other.

Table A7.2. Contrasts used in testing all possible pairs of treatments from a four-treatment experiment.

	A versus B	A versus C	A versus D	B versus C	B versus D	C versus D
A	+1	+1	+1			
B	−1			+1	+1	
C		−1		−1		+1
D			−1		−1	−1
$\sum c =$	0	0	0	0	0	0

This set satisfies rules **1** and **3** but violates rule **2**: there are six contrasts, but only 3 d.f. We could resolve that difficulty by leaving out three of the columns. However, you will find that when you apply rule **4**, column 1 is not orthogonal with columns 2, 3, 4 or 5: there is no way of making an orthogonal set by selecting three out of the six columns. If you begin by contrasting A and B, then you must continue by contrasting *either* A and B versus C (leaving as the final contrast A+B+C versus D) *or* A+B versus C+D, leaving C versus D as the third contrast.

If you play a little longer you will find that, with four treatments, there are only three orthogonal *combinations* available,[a] although, by deciding which treatments you call A, B, C and D there are substantially more *permutations*. The sets available are given in Table A7.3. With larger numbers of treatments there are more choices of orthogonal sets.

Table A7.3. Orthogonal sets of contrasts available for an experiment with four treatments (A, B, C and D). Alternative versions of sets 1 and 3 can be obtained by relabelling the treatments.

	Set 1			Set 2			Set 3		
	Contrasts:			Contrasts:			Contrasts:		
	1	2	3	1	2	3	1	2	3
A	3	0	0	1	1	1	1	1	0
B	−1	2	0	1	−1	−1	1	−1	0
C	−1	−1	1	−1	1	−1	−1	0	1
D	−1	−1	−1	−1	−1	1	−1	0	−1

It is not essential to divide the treatment s.s. into single d.f. components and sometimes it is not worth bothering. For example, if we have a set of treatment means as in Table A7.4 it may be sensible to evaluate the following contrasts:

[a] That is, tests that ask questions of the 'this versus that' type. You can ask other sorts of questions, including whether there is a linear, quadratic or cubic trend in the data. The coefficients for these are given below.

Table A7.4. Treatment means from a hypothetical experiment.

Treatment					
A	B	C	D	E	SEM
137.1	135.8	146.3	145.9	148.8	3.1

1. (A + B) versus (C, D and E);
2. A versus B;
3. amongst C, D and E.

Contrasts **1** and **2** can be quickly calculated from treatment totals and the third s.s. (with 2 d.f.) can be found by difference from the treatment s.s.

Orthogonal Polynomials for Regression Analysis

Finally, if you are now persuaded that polynomials give you an easy method of finding s.s. (with a single d.f.), you may like to have a list (Table A7.5) of the polynomials that can be used to derive linear, quadratic and cubic component s.s. for regressions with up to six *equally spaced* treatments.

Table A7.5. Polynomials for calculating linear (l), quadratic (q) and cubic (c) regression s.s. when treatment doses are equally spaced.

	Number of treatments											
	3		4			5			6		
	l	q	l	q	c	l	q	c	l	q	c
	1	+1	+3	+1	+1	+2	+2	+1	+5	+5	+5
	0	−2	+1	−1	−3	+1	−1	−2	+3	−1	−7
	−1	+1	−1	−1	+3	0	−2	0	+1	+4	−4
			−3	+1	−1	−1	−1	+2	−1	−4	+4
						−2	+2	−1	−3	−1	+7
									−5	+5	−5
$\Sigma c^2 =$	2	6	20	4	20	10	14	10	70	84	180

Thus, if you have four treatment totals, obtained by applying equally spaced doses, as follows:

26.3 29.9 34.1 37.0 (each being the total of **five replicates**)

the linear regression s.s. is:

$$\frac{\{+3(26.3) + 1(29.9) - 1(34.1) - 3(37.0)\}^2}{20 \times 5} = 13.1769.$$

The linear regression coefficient is obtained by leaving out the squaring operation in the top line:

$$b = \frac{3(26.3) + 29.9 - 34.1 - 3(37.0)}{20 \times 5} = -0.363,$$

but now the *scale* and the *sign* of the coefficients employed do matter. Assuming that our treatments were described in *ascending* order of doses, we have used coefficients that alter the input variable by **−2** for each *increment* in dose. We must therefore divide b by −2 to discover that the response increases by +0.1815 for each additional unit of input. This b value may need further re-scaling, depending on the nature of our doses, before we have it in units that are suitable for publication.

Appendix 8

Answers to Exercise 3.1

This is an example in which 'peeling off' treatments one at a time is probably the best strategy. We would start by asking whether the unsupplemented diet differs from the average of the rest (it obviously does). Then our attention would be caught by the effect of omitting riboflavin. Having shown that this is also a significant effect, we might wonder whether the omission of niacin has produced a significant drop in growth rate and move on to test that. This process would continue until we have a remaining group of treatments amongst which there cannot be any significant differences because the ratio of the *sum of squares* for that remaining group to the *mean square* for error is smaller than the 5% F-ratio for 1 and [error] d.f.

We were not given the original ANOVA and so cannot complete all the calculations, but part of the ANOVA framework would be as in Table A8.1.

Table A8.1. Subdivision of the treatments s.s. for Exercise 3.1.

Source	d.f.	
Treatments	7	
Unsupplemented versus rest		1
Omitting riboflavin versus other supplements		1
Omitting niacin versus remaining five supplements		1
Amongst remaining five supplements		4

Note that the first three questions can be answered using orthogonal polynomials (if we knew the number of replicates) but can also be tested with Student's t (which is listed in Appendix 26), since they are contrasts between two quantities. Thus:

Unsupplemented mean = 14.4 g day^{-1}

Mean of the rest = 17.4 g day^{-1}

Difference = 3.0
SED = $\sqrt{(1 + 1/7)(0.63)} = 0.673$
$t = 3.0/0.673 = 4.46^{**}$
(d.f. unknown, but ≥ 7)

Similarly:

Mean omitting niacin = 16.2 g day^{-1}	Difference = 2.1
	SED = $\sqrt{(1 + 1/5)(0.63)}$ = 0.690
Mean of remaining five = 18.3 g day^{-1}	t = 2.1/0.690 = 3.04*

Notice that a *multiple range test* applied to these data would *not* have revealed that the omission of niacin caused a significant depression in growth rate.

The Strategy of the Design

We should allow for the possibility that the unsupplemented organic diet might be deficient in more than one of the B vitamins (it might also be deficient in vitamin A; but if the birds are in a green environment they will derive vitamin A from carotene). Adding the vitamins one at a time will not do much good if there is a deficiency of more than one. Adding the vitamins in all possible combinations requires $2^5 = 32$ treatments and is impracticable. The device of feeding the complete mixture and then omitting each vitamin in turn will reveal whether a supplement of one or more of them is needed for normal growth.

Answers to Exercise 3.2

The sensible initial contrasts would be:

Untreated versus rest	(1 d.f.)
NaOH versus NH_4OH	(1 d.f.)
Alkali levels	(2 d.f.)
Levels × alkali type	(2 d.f.)

You would probably find that all these contrasts proved significant or were dangerously close to being significant.

The proper analysis is then one that looks at the dose–response relationship separately for NaOH and for NH_4OH. Test for linear and curvilinear regression (remembering that the untreated straw is a zero dose and is an important point on the line or curve). Some would argue that the control treatment in this design should have received double replication to allow for its subsequent use in two 'independent' regressions, but there is little merit in this argument. If you had used three replications for each of the treated straws and six for the control, there would be no point in associating a particular three control values with one regression line and the remaining three with the other. If, on the other hand, you use the mean of six control values in both regression analyses you have simply given extra replication to the region of the curve which is of least interest. Technically, the two regression curves are not independent because they share a common estimate of response to the zero dose; but independence is of no great importance in interpreting this experiment.

We do not have the data to complete the ANOVA for Exercise 3.2 but we can easily calculate the two regressions from the treatment means supplied. Since the doses are evenly spaced, the simplest method is to use the orthogonal polynomials given in Appendix 7 for a four-point regression (−3, −1, +1 and +3).

Set out the data as follows:

	x (g NaOH (kg straw)$^{-1}$)	y (OMD (g kg^{-1}) in straw)
	0	421
	20	480
	40	532
	60	599
Totals	120	2032
Means	30	508

OMD, organic matter digestibility.

The linear regression coefficient for the NaOH treatments will be:

$$b = \frac{-3(421) - 1(480) + 1(532) + 3(599)}{20^{a}}$$

= 29.3 increase in OMD units for each 10 g kg^{-1} [b] increase in NaOH

or rescaled $b = 2.93$ OMD units for each 1 g kg^{-1} NaOH.

The linear regression equation is:

$$y = y + b(x - \bar{x}) = 508 + 2.93(x - 30) = \mathbf{420.1 + 2.93}\boldsymbol{x}.$$

You can now perform a similar calculation to arrive at the linear regression equation for the NH_4OH treatment, which is:

$$\boldsymbol{y} = \mathbf{418.5 + 1.825}\boldsymbol{x}.$$

A glance at Fig. A8.1 should convince you that it is not worth pursuing the calculation of curvilinear regressions in this case.

[a] This 20 is the sum of squares of the coefficients: there is no 'n' to enter, because the numbers above are means, not totals. If you want to derive the regression sum of squares from means, you should square the top line and then *multiply* it by the number of replicates in each mean. If this seems strange, consider the equation:

$$\frac{(\text{total A} - \text{total B})^2}{\Sigma c^2.n} = \frac{n\{(\text{total A})/n - (\text{total B})/n\}^2}{\Sigma c^2}.$$

[b] This number is 10 g kg^{-1} because we have used coefficients that conveniently avoid fractional numbers: thus we have set 2 on the coefficient scale to represent the dose interval of 20 g kg^{-1}.

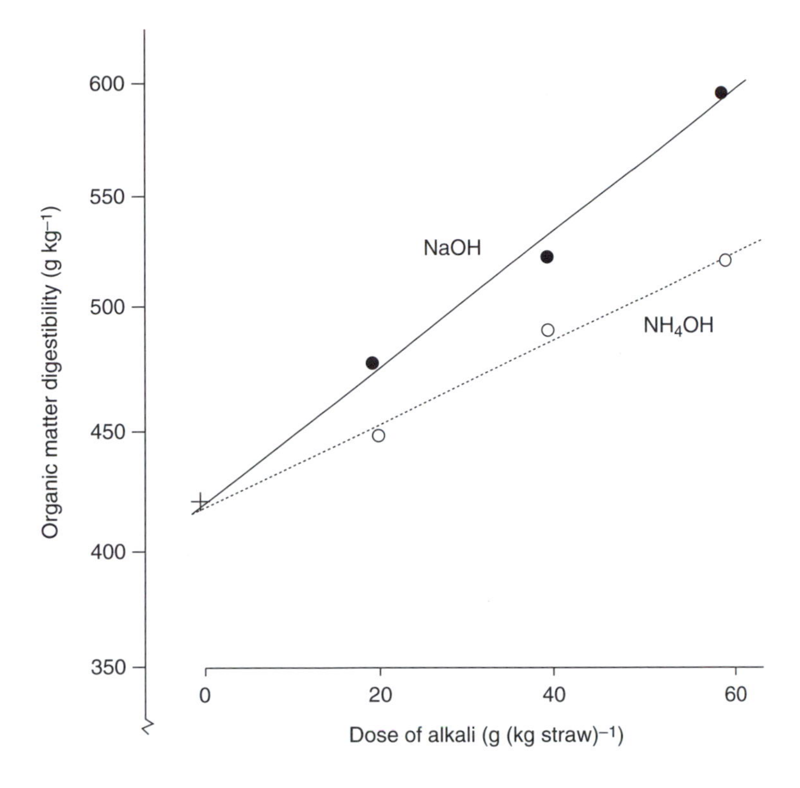

Fig. A8.1. The linear regression of OMD on dose of alkali from the data of Exercise 3.2.

Appendix 9

Answer to Exercise 4.1

$n = 8(12)^2/8^2 = 18$ pigs per treatment.

As there are six diets to be tested, the *total number of pigs* required for the experiment will be *108*.

Answer to Part 1 of Exercise 4.2

$n = 2(t_1 + t_2)^2 \,.\, (\mathrm{CV})^2/d\%^2$

CV = 6; $d\%$ = 10; t_1 and t_2 depend on error d.f. (as yet unknown).

Begin by *guessing* that $n = 10$ (or any other number you fancy). Ten replicates would require 40 sheep-periods to compare four roughages, and the ANOVA framework could be as simple as Table A9.1.

Table A9.1. Framework for ANOVA of ten replicates of four treatments with no blocking or period effects to isolate.

Source	d.f.
Diets	3
Error	36
Total	39

Some experimenters would prefer a design that removes average differences between sheep and between periods, but this gives away d.f. without necessarily bringing any benefit (see the discussion of Latin squares in Chapter 5).

For 36 d.f. and $P = 0.05$, $t_1 = 2.029$.
The probability for t_2 $= 2(1 - p)$, where p is given as 0.8
$= 2(1 - 0.8) = 0.4$.
For 36 d.f. and $P = 0.40$, $t_2 = 0.852$.
First estimate of n $= 2(2.029 + 0.852)^2 \,.\, 6^2/10^2 = 5.98$.

For $n = 6$ replicates and four treatments, error d.f. = 20
For 20 d.f. and $P = 0.05$, $t_1 = 2.086$
For 20 d.f. and $P = 0.40$, $t_2 = 0.860$
$t_1 + t_2 = 2.946$.
Second estimate of n $= 2(2.946)^2 \,.\, 6^2/10^2 = 6.25$.

Six replicates are not quite sufficient: *check* that seven will do:

For seven replicates and four treatments, error d.f. = 24

$t_1 = 2.064$, $t_2 = 0.857$, $n = 2(2.921)^2 \,.\, 36/100 = 6.14$.

The experiment will need seven replicates (i.e. 28 sheep-periods in total) to have an 80% chance of detecting differences of 6% as significant at $P = 0.05$.

Answer to Part 2 of Exercise 4.2

To calculate the number of replicates required for a given LSD we inverted the equation

$$\text{LSD} = t \,.\, \sqrt{(2s^2/n)}$$

and so obtained

$$n = 2\,t^2 \,.\, s^2/d^2.$$

Similarly, if we want to know how many replicates will be required to achieve a given SEM, we invert the equation

$$\text{SEM} = \sqrt{(s^2/n)}$$

to obtain

$$n = s^2/(\text{SEM})^2$$

or, converting both the right-hand-side quantities to percentages of the general mean,

$$n = (\text{CV})^2/(\text{SEM}\%)^2.$$

In the problem set, CV = 6%, SEM% = 2/50 = 4% and therefore $n = 6^2/4^2 = 2.25$.

Two replicates would not be quite sufficient to give the precision asked for. It might be helpful to calculate that, given a CV of 6%

for $n = 2$, SEM% = $\sqrt{(6^2/2)} = 4.2\%$ of the general mean;

for $n = 3$, SEM% = $\sqrt{(6^2/3)} = 3.5\%$ of the general mean.

Answer to Exercise 4.3

Given: $d\%$ = 5; probability of success = 0.5; error d.f. = 3(n − 1), we calculate:

	CV (%)					
	6	9	12	15	20	25
d%	Number of replicates required					
5	12	26	46	71	125	194

Appendix 10

Answers to Exercise 5.1 Using a Pocket Calculator

First calculate the *period totals* (not given in Exercise 5.1) by summing the values labelled with Roman numerals to obtain:

Period	Period totals
I	216.5
II	218.7
III	217.2
IV	212.6
Total	865.0

1. Correction factor (CF) = $865.0^2/16 = 46\,764.0625$.

2. Period s.s. $= \dfrac{\Sigma(216.5^2 + \ldots\ 212.6^2)}{4} - \text{CF} = \mathbf{5.0725}$.

3. Groups s.s. $= \dfrac{\Sigma(217.6^2 + \ldots\ 211.6^2)}{4} - \text{CF} = \mathbf{9.0275}$.

4. Treatments s.s. $= \dfrac{\Sigma(218.7^2 + \ldots\ 212.6^2)}{4} - \text{CF} = \mathbf{6.3725}$.

5. Total s.s. is given as **22.6575** (= $\Sigma\{(\text{plot values})^2\} - \text{CF}$).
Error s.s., by difference,

$$= 22.6575 - (5.0725 + 9.0275 + 6.3725) = \mathbf{2.185}.$$

6. Enter your results in a table (see Table A10.1).

The *F*-ratio for treatments tells us that the diets had a significant effect on egg weight, and inspection of the means shows that egg weight got smaller as the concentration of molasses in the diet increased. The figures therefore justify a conclusion that the inclusion of molasses depresses egg weight. (If you are curious as to *why* molasses depressed egg weight in this experiment, you may like to reflect that molasses, because of its water content, has a much lower energy density than the basal diet).

Table A10.1. Completed ANOVA of the Latin square design.

Source	d.f.	s.s.	m.s.	F
Periods	3	5.0725	1.69083	4.64
Groups	3	9.0275	3.00917	8.26*
Treatments	3	6.3725	2.12417	5.83*
Error	6	2.1850	0.36417	
Total	15	22.6575		
	SEM = 0.302 g		LSD (P = 0.05) = 1.044 g	

The next question you are invited to tackle is 'What is a safe level of molasses – supposing that *any* reduction in egg weight is unacceptable?' You may well think that the question is unrealistic, because:

1. a *small* change in egg size would not be noticed by the farmer (and might not be of any consequence to the customer who buys the eggs);
2. a more sensible question might be 'What is the most *profitable* level of molasses inclusion, taking into account the effect on egg weight?' (this question is answered below).

Nevertheless the problem has been deliberately posed in the form 'What is a safe level ...' because it is not uncommon to have to face a question which amounts to asking 'Up to what dose level is there *zero* response?'

Since the LSD = 1.044 g, it is clear that the 70 and 140 g kg^{-1} levels of molasses have not caused a *significant* reduction in egg weight (but 210 g kg^{-1} did). Here is a *trap for the unwary*! You may be tempted to conclude that 140 g kg^{-1} is a safe level of molasses, but how confident are you that such a level would cause *no reduction* in egg weight? The difference between the 0 and the 140 level is 0.88 g, the standard error of this difference is 0.427 (SEM. $\sqrt{2}$) and $t = 0.88/0.427 = 2.06$, for which $P < 0.10$. The odds are greater than 10:1 that the 140 g kg^{-1} diet really did *cause* a reduction in egg weight.

The best estimate of the relationship between dietary molasses and egg weight will be obtained if we recognize this as a dose–response experiment (which it is) and conduct a regression analysis.

The doses of molasses are equally spaced (increments of 70 g kg^{-1}) and so we can use orthogonal polynomials (see Appendix 7) and the treatment totals given in Exercise 5.1 to calculate the regression s.s.

$$\text{Linear regression s.s.} = \frac{\{3(218.7) + (218.5) - (215.2) - 3(212.6)\}^2 = 5.8320}{20 \times 4}.$$

We could also calculate the quadratic and cubic components, but looking at the treatment s.s. (6.3725) and the linear regression component (5.8320) we can see that the bit left over (0.54) is much too small to be worth splitting, since the error m.s. = 0.36. Our final subdivision of the treatment s.s. is set out in Table A10.2.

Table A10.2. Regression analysis within the Latin square design.

Source	d.f.	s.s.	m.s.	*F*
Linear regression	1	5.8320	5.83200	16.01**
Deviations from regression	2	0.5405	0.27075	
Total for treatments	3	6.3725	2.12427	5.83*
Error	6	2.1850	0.36417	

A linear regression is a good fit to these data and there is no indication that a curve (or any other more complex model) would be justified. Some further discussion of possible models that might be investigated in other cases will be found in Chapter 9. We can now draw a graph (Fig. A10.1) to represent our conclusions.

The linear regression coefficient, *b*, is obtained by using the orthogonal polynomials given above, but omitting the squaring operation used to derive the s.s.:

$$b\text{ (coded)} = \frac{3(218.7) + (218.5) - (215.2) - 3(212.6)}{20 \times 4} = 0.27.$$

Egg weight increases by 0.27 g for each coded unit of input. In the units (orthogonal polynomials) we have used −2 represents +70 g kg^{-1} molasses in the diet and so we calculate:

b (decoded) = 0.27(−2/70) = −0.0077 g egg weight for each 1 g kg^{-1} extra molasses in the diet.

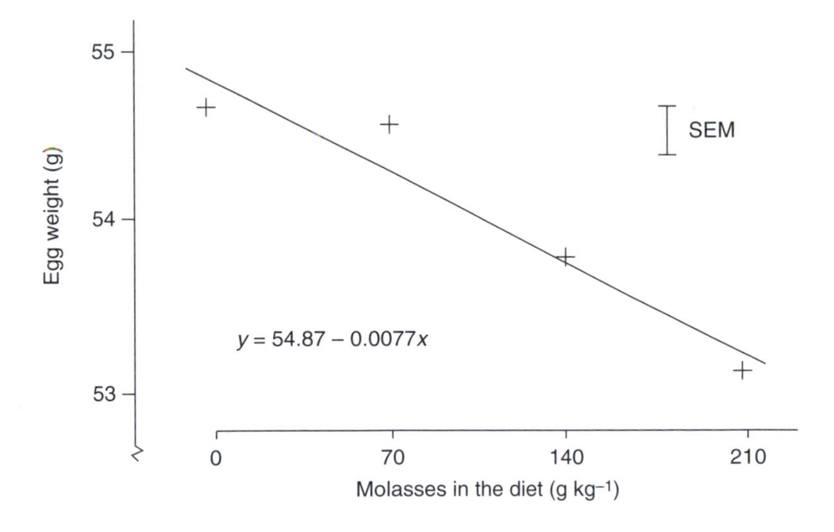

Fig. A10.1. The relationship between rate of molasses inclusion and egg weight in the experiment reported in Exercise 5.1.

$\bar{y} = 865.0/16 = 54.06$ g; $\bar{x} = 105$ g kg^{-1} molasses.

The regression equation is:

$$y = \bar{y} + b(x - \bar{x}) = 54.06 - 0.0077(x - 105) = \mathbf{54.87 - 0.0077}\boldsymbol{x}.$$

Since the data support the hypothesis of a (negative) linear relationship between egg weight and molasses inclusion rate throughout the *whole range* from 0 to 210 g kg^{-1}, we must conclude that the only safe level that has no effect on egg weight is 0 g kg^{-1}.

If you suspect that a 'bent-stick' model, with zero effect of molasses up to 70 g kg^{-1}, but a linear decline thereafter, would be a better representation of the data, you should read Chapter 9, where this hypothesis is discussed.

Optimizing profit

Accepting the straight line in Fig. A10.1 as the best representation of the results of this experiment, how do we determine a *most profitable dose*, using the 'balance of probability' argument discussed in Chapter 3, which says that we can opt for the dose that is most likely to give the highest profit, even though the odds in favour of it being more profitable than other doses may not be very high?

Assuming the value of eggs to be directly proportional to their weight (which is seldom the case) and the cost of the diet to be proportional to the rate of dilution with molasses (a more likely assumption), the maximum profit will be found either at the lowest dose (zero molasses) or at the highest dose (or possibly at some still higher dose which we have not tested). There cannot be an intermediate point of maximum profit for a straight-line function. If there is a more complex relationship between the size of eggs and their market value, we need to replace the left-hand scale (called the ordinate) by the average *value* of the eggs produced by each treatment. This may be a complex calculation requiring an algorithm involving the grading pattern represented by these means and the value of eggs falling into different grade sizes.

Appendix 11

Answers to Exercise 5.1 Using SAS

Before beginning, it is useful to set the page size to 54 lines and the line size to 76 characters so that (provided your font size in SAS is 12 cpi) your output will fit nicely on a page when printed. This is done by entering:

```
OPTIONS PS=54 LS=76;
```

Variable names are assigned and the data are read in as shown in Table A11.1. To plot the data for visual assessment enter:

```
PROC PLOT; PLOT EGGWT*DIET; RUN;
```

To complete the ANOVA, enter the commands shown below. Note that type 3 s.s. have been specified. As the study is a dose–response trial with equally

Table A11.1. The input array for the data from the Latin square in Exercise 5.1.

```
DATA AE51;
INPUT GROUP $ 1 PERIOD 3 DIET 5–7 EGGWT 9–12;
CARDS;
A 2 0     55.4
A 3 70    55.1
A 4 140   53.6
A 1 210   53.5
B 4 0     55.0
B 1 70    56.1
B 2 140   54.8
B 3 210   53.9
C 3 0     55.2
C 4 70    52.9
C 1 140   53.8
C 2 210   54.1
D 1 0     53.1
D 2 70    54.4
D 3 140   53.0
D 4 210   51.1
RUN;
```

spaced treatments, `CONTRAST` statements have been used so that SAS makes the specified orthogonal contrasts to test for a linear or quadratic relationship between diet and egg weight. The diet means and a comparison of those means using the LSD at $P = 0.05$ have also been asked for, as well as least squares means for diets, with the appropriate SEM.

```
PROC GLM;
CLASSES GROUP PERIOD DIET;
MODEL EGGWT=GROUP PERIOD DIET/SS3;
CONTRAST 'LINEAR' DIET 3 1 -1 -3;
CONTRAST 'QUADRATIC' DIET 1 -1 -1 1;
MEANS DIET/LSD;
LSMEANS DIET/STDERR;
QUIT;
```

You can also test for linear responses and derive the equation describing the linear regression using '`PROC GLM;`' and the '`SOLUTION`' option, as follows:

```
PROC GLM;
MODEL EGGWT=DIET/SOLUTION;
QUIT;
```

Note that diet is no longer specified as a discrete variable. The quadratic response can be tested by adding '`DIET*DIET`' as an independent variable to this model.

Appendix 12

Example ANOVA of a Balanced Latin Square Design

The advantages and disadvantages of balanced Latin squares are discussed in Chapter 5. This worked example illustrates the method for completing the calculations using a pocket calculator.

The data in Table A12.1 represent yields of six dairy cows allocated to three treatments in a pair of balanced LS. The treatments, A, B and C were increasing amounts of concentrate feed supplied as a supplement to a forage diet which was fed *ad libitum*. Note that two is the minimum number of squares required to achieve balance when there are only three treatments. In this design each treatment follows each other treatment twice.

Table A12.1. Yields (kg day^{-1}) of dairy cows allocated to three treatments (A, B and C) in a balanced pair of Latin squares.

	Square 1				Square 2			
	Cow				Cow			
Period	1	2	3	Period totals	4	5	6	Period totals
I	**A**20.2	**C**32.8	**B**30.6	83.6	**A**27.3	**B**25.7	**C**29.5	82.5
II	**B**18.3	**A**25.2	**C**27.3	70.8	**C**25.8	**A**19.8	**B**23.9	69.5
III	**C**16.9	**B**18.6	**A**20.4	55.9	**B**21.4	**C**19.6	**A**14.8	55.8
Cow totals	55.4	76.6	78.3	210.3	74.5	65.1	68.2	207.8
							Grand total =	418.1

The first step is to set out a table giving the structure of the ANOVA which will help in identifying the quantities that have to be calculated. You will find this structure in Table A12.3. In the experiment described, the first three cows were put on trial several weeks before cows 4–6 and so the period effects in square 1 were not coincident with period effects in square 2. For this reason we must allocate 4 d.f. (2 d.f. from each square) to periods *within squares*. Had all six cows been started at the same time we could have assumed common period effects for the two squares and allocated only 2 d.f., thus assigning another 2 d.f. to error.

The initial calculations are straightforward.

$$\mathrm{CF} = 418.1^2/18 = 9711.5339.$$

$$\text{Total s.s.} = (20.2^2 + 32.8^2 + \ldots + 14.8^2) - \mathrm{CF} = 437.736.$$

$$\text{Cows s.s.} = \tfrac{1}{3}(55.4^2 + 76.6^2 + \ldots + 68.2^2) - \mathrm{CF} = 124.169.$$

$$\text{Periods s.s.} = \tfrac{1}{3}(83.6^2 + 70.8^2 + 55.9^2) - (210.3^2/9) + \tfrac{1}{3}(82.5^2 + 69.5^2 + 55.8^2) - (207.8^2/9) = 246.969.$$

Now we must calculate the treatment totals (T), by summing the appropriate plots from Table A12.1. These numbers are shown in Table A12.2. We also need the residual totals (R) which are defined as the sum of the plot yields in the period immediately following each treatment. Thus R for treatment A = 18.3 + 18.6 + 25.8 + 19.6 = 82.3. We also need the final totals (F) which are the sum of the column (= cow) totals in which each treatment is the final one. F for treatment A = 78.3 + 68.2 = 146.5.

Table A12.2. Quantities needed for the ANOVA (T, R, F, t, r and u are defined in the text).

Treatment	T	R	F	t	r	u
A	127.7	82.3	146.5	−56.3	−28.9	6.1
B	138.5	78.8	151.1	−8.2	−14.2	−11.6
C	151.9	90.9	120.5	64.5	43.1	5.5
Totals	418.1	252.0	418.1	0	0	0

$$\text{Treatment s.s. unadjusted} = \{(127.7^2 + 138.5^2 + 151.9^2)/6\} - \mathrm{CF} = 48.991.$$

Now we must define a number of other symbols which will appear below:

m = number of Latin squares used (2 in this case);
n = number of treatments (3 in this case);
G = the grand total (418.1 in this case);
P_{I} = the total for period I (166.1 in this case);
(note, as a check, that $\sum R = G - P_{\mathrm{I}}$).

Also enter in Table A12.2 the estimated direct effects of treatment (t), the estimated residual effects (r) and the unadjusted residual effects (u) for each treatment, calculated as follows:

$$t = (n^2 - n - 1)T + nR + F + (P_{\mathrm{I}} - nG)$$
$$= 5T + 3R + F + (166.1 - 1254.3).$$

$$r = nT + n^2R + nF + (nP_{\mathrm{I}} - [n + 2]G)$$

$$= 3T + 9R + 3F + (498.3 - 2090.5).$$

$$u = R + G - nT.$$

Check that these quantities sum to zero in every case.

The adjusted treatment s.s is $\sum t^2/\{mn(n^2 - n - 1)(n^2 - n - 2)\}$

$$= \{(-56.3)^2 + (-8.2)^2 + (64.5)^2\}/120 = 61.643.$$

The adjusted residual s.s. is $\sum r^2/\{mn^3\,(n^2 - n - 2)\}$

$$= \{(-28.9)^2 + (-14.2)^2 + (43.1)^2\}/216 = 13.400.$$

The unadjusted residual s.s. is $\sum u^2/\{mn^3\,(n^2 - n - 1)\}$

$$= \{(6.1)^2 + (-11.6)^2 + (5.5)^2\}/270 = 0.748.$$

Check that (unadjusted treatment s.s. + adjusted residual s.s.) = (adjusted treatment s.s. + unadjusted residual s.s.). We can now enter all these quantities in Table A12.3.

Table A12.3. ANOVA table for a pair of balanced 3 × 3 Latin squares.

Source	d.f.		s.s.	m.s.	*F*
Cows	5		124.169	24.834	23.61^{**}
Periods within squares	4		246.969	61.742	58.70^{***}
Diets (unadjusted)	2		48.991		
Residual effects (adjusted)	2		13.400	6.700	6.37 $P<0.10$
[Diets (adjusted)		2	61.643	30.821	29.30^{**}]
[Residual effects (unadjusted)		2	0.748		]
Error	4		4.207	1.05175	
Total	17		437.736		

The adjusted mean for each treatment is found by adding to the general mean (i.e. 418.1/18) the quantity $t/\{mn(n^2 - n - 2)\} = t/24$ in this case.

The SE of an adjusted treatment mean is:

$$\sqrt{[s^2/r][(n^2 - n - 1)/(n^2 - n - 2)]}$$

$$= \sqrt{[1.05175/6][5/4]} = 0.468 \text{ kg day}^{-1}.$$

Thus the design, although using only six cows, was quite effective in detecting differences of about 10% between dietary treatments. The residual effects of treatment carrying over from one period to the next were not significant at $P = 0.05$ but were nevertheless six times the size of the error term and therefore worth isolating. As with the discussion of blocking effects in Chapter 2, the 'significance' of adjusted residual effects is much less important than their size.

Table A12.4. Summary of results. Mean yields (kg day^{-1}) adjusted for residual effects of treatment carried over from the preceding period.

Treatment A	20.88
Treatment B	22.89
Treatment C	25.92
SEM	0.468

Appendix 13

Answers to Exercise 6.1

Finding an appropriate SEM

The important thing to recognize in this exercise is that the pens (each containing ten chicks) are the experimental units, *not* the individual chickens. We must calculate a *between-pen error* with which to test dietary effects. This is most easily done by analysing the ten pen means. Although these means differ slightly in their precision, because some are based on ten chicks and others on only nine survivors, we would not usually pay any attention to that slight difference, but treat all ten means as equally good estimates of chick growth in the pen in question.

The ANOVA of the ten pen means is given in Table A13.1. This involves no novel computations. The method of computing sums of squares for treatments (vitamin A levels) is given in Appendix 3. Notice that there is no blocks s.s., because pens were allocated to treatments at random, not in two blocks of five. Presumably the experimenter judged that allocating 1 d.f. for blocks was not likely to isolate a useful chunk of variation in this case.

Table A13.1. ANOVA of the pen means listed in Exercise 6.1.

Source	d.f.	s.s.	m.s.	*F*
Vitamin A levels	4	47,913.41	11,978.35	42.4^{***}
Error	5	1414.08	282.82	
Total (amongst pens)	9	49,327.49		

$$\text{SEM} = \sqrt{(282.82/2)} = 11.9 \text{ g}$$

So, the *appropriate* SEM for each treatment is 11.9 g.

The full ANOVA

A full ANOVA showing within-pen as well as between-pen variation can be computed by using the 98 individual chick weights. In addition to the numbers listed in Exercise 6.1, we shall need to write down the treatment totals, as in Table A13.2.

Table A13.2. The treatment totals.

	Vitamin A					
	0	500	1000	2000	4000	Grand total
Treatment total	4765	7305	7333	8847	8749	36999
(n)	(19)	(20)	(19)	(20)	(20)	(98)

The treatment s.s. can now be computed as:

$$\text{Treatment s.s.} = \frac{4765^2 + 7333^2}{19} + \frac{7305^2 + 8847^2 + 8749^2}{20} - \frac{36999^2}{98}$$

$$= \mathbf{465{,}402.98}.$$

$$\text{Pens s.s.} = \frac{2352^2 + 3609^2}{9} + \frac{2413^2 + 3535^2 + \ldots + 4306^2}{10} - \frac{36999^2}{98}$$

$$= \mathbf{479{,}228.34}.$$

Error (a), between pens, is found by difference:

Between-pen error s.s. = 479,228.34 − 465,402.98 = **13,825.36**.

You should notice that the three s.s. we have just calculated are approximately ten times the quantities shown in Table A13.1, which were based on pen means. If there had been ten chicks in *every* pen, variances calculated from individual chick weights would have been *exactly* 10 times the variances calculated from pen means (and, in that case, F-ratios would have been identical for the two analyses).

The full ANOVA is given in Table A13.3. The total s.s. for chicks can be calculated in the usual way and error (b) can be found by difference. However, in this case, because the s.s. within each pen has already been computed (see Exercise 6.1) you could obtain the error s.s. directly by adding the ten s.s. given in Exercise 6.1. This would save you the bother of totting up the 98 squares of individual chick weights to obtain the total s.s.

Examining Table A13.3, you will notice that error (a) is nearly twice as big as error (b). This is often the case in chick experiments of this sort and means that chick growth, in addition to its basic variability, has been influenced by factors associated with the pens in this trial. An SEM calculated from the analysis of individual chick variation *within pens* would be too small for a valid test of treatment effects, because it would leave out this element of pen-to-pen variation. Notice that the CV for chicks (within pens) = 10.2% (($\sqrt{1490}$)/377.5), but the CV for pens of ten chicks is 4.45%, considerably larger than the 3.2% (10.2%/$\sqrt{10}$) expected on the basis of within-pen variation alone.

Table A13.3. ANOVA of individual chick weights from Exercise 6.1.

Source	d.f.	s.s.	m.s.	F
Vitamin A levels	4	465,402.98	116,350.75	42.1***
Error (a) (amongst pens)	5	13,825.36	2765.07	
Pens	9	479,228.34		
Error (b) (amongst chicks)	88	131,148.79	1490.33	
Total (chicks)	97	610,378.33		

A linear regression analysis can be carried out in the context of either ANOVA. Usually, we would be content to use the (unweighted) treatment means, which saves the bother of having to weight values in the regression according to the number of chicks contributing to each mean. If the number of chicks in each group varied more dramatically, a properly weighted analysis would be preferable.

Since the treatments are *not equally spaced* in this case (and even if we convert to a logarithmic scale, the control diet does not fit on an additive scale with the other four treatments), the orthogonal polynomials in Appendix 7 are not applicable and we shall need to calculate suitable coefficients for a linear regression analysis, as shown in Table A13.4. The ANOVA in Table A13.1 can now be extended to include the variance components listed in Table A13.5. The linear regression of chick weight on dose of vitamin A is illustrated in Fig. A13.1.

Although a good deal of the variation between treatments is accounted for by a straight line, the deviations from linear regression are significantly larger than error and so a different model is called for. A curvilinear model might be more suitable, but we should do well to choose a model that reaches a plateau. A

Table A13.4. Calculation of the linear regression.

	x	$(x - \bar{x})$	c $\{= (x - \bar{x})/500\}$	Treatment totals (n = **2**)
	0	−1500	−3	502.6
	500	−1000	−2	730.5
	1000	−500	−1	773.4
	2000	+ 500	+1	884.7
	4000	+2500	+5	874.9
$\Sigma =$	7500	0	0	3766.1
$\bar{x} =$	1500		$\Sigma c^2 = 40$	

Linear regression s.s. =

$$\frac{\{-3(502.6) - 2(730.5) - 1(773.4) + 1(884.7) + 5(874.9)\}^2}{40 \times \mathbf{2}} = 28{,}766.11$$

Table A13.5. The ANOVA of regression components.

Source	d.f.	s.s	m.s.	*F*
Linear regression	1	28,766.11	28,766.11	101.7***
Deviations	3	19,147.30	6382.43	22.6**
Vitamin A levels	4	47,913.41	11,978.35	42.4**
Error	5	1414.08	282.82	

quadratic equation would imply a reduction in growth rate when the dietary vitamin A level is slightly in excess of the chicks' requirement and the data do not justify such an implication.

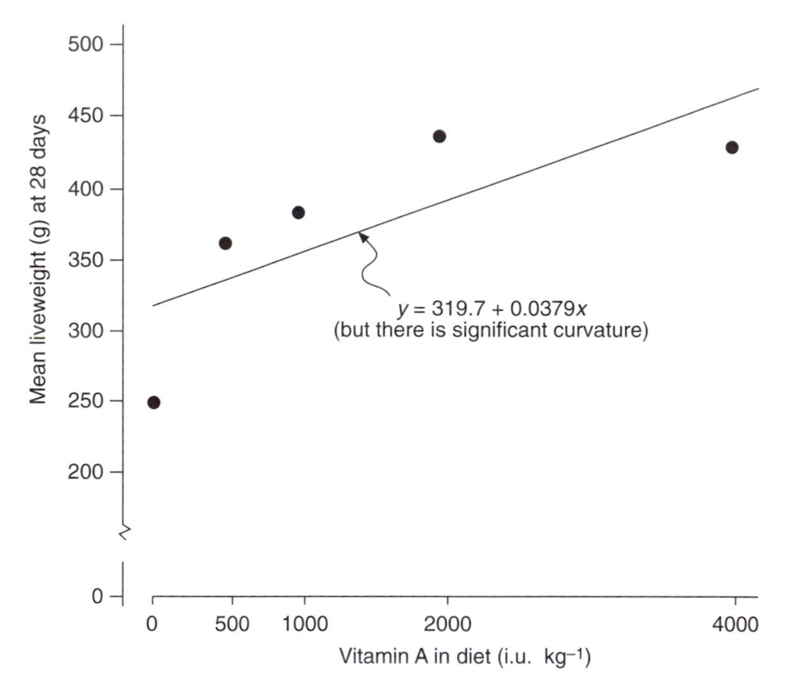

Fig. A13.1. Regression of weight of 28-day-old chicks on dietary concentration of vitamin A.

Was it necessary to weigh each chick?

It was not necessary to weigh chicks individually in this experiment, but the quality of the *feed intake* data for each pen might be marginally improved by knowing the weights of (and so being able to impute feed intakes to) the chicks that died.

Appendix 14

Answers to Exercise 7.1

A method for calculating an interaction s.s. is given in Appendix 5. This would involve, in this case, setting out two-way tables of totals for Lys × Met, Lys × Cys and Met × Cys, which is laborious. However, because this is a 2 × 2 × 2 (or 2^3) design there will be *1 d.f.* for each main effect and each interaction (see Table A14.2) and this allows us to find all the s.s. by setting out suitable orthogonal polynomials as in Table A14.1.

The main effect of lysine is found by putting +1 against all the diets that included lysine and −1 against the diets without added lysine, as in column 1 of the matrix.

The lysine s.s is then calculated as:

$$\frac{\{-1(920.3) + 1(909.7) - 1(970.2) - 1(979.0) + 1(1000.8) + 1(992.0) - 1(960.1) + 1(994.2)\}^2}{\mathbf{8 \times 6}}$$

$$= \mathbf{93.8002}.$$

Table A14.1. Polynomials for finding main effect and interaction sums of squares.

Diet	Diet totals (n = **6**)	Main effects			Interactions			
		Contrast:			Contrast:			
		L	M	C	LM	LC	MC	LMC
0	920.3	−1	−1	−1	+1	+1	+1	−1
L	909.7	+1	−1	−1	−1	−1	+1	+1
M	970.2	−1	+1	−1	−1	+1	−1	+1
C	979.0	−1	−1	+1	+1	−1	−1	+1
LM	1000.8	+1	+1	−1	+1	−1	−1	−1
LC	992.0	+1	−1	+1	−1	+1	−1	−1
MC	960.1	−1	+1	+1	−1	−1	+1	−1
LMC	994.2	+1	+1	+1	+1	+1	+1	+1
	$\sum c$ =	0	0	0	0	0	0	0
	$\sum c^2$ =	**8**	8	8	8	8	8	8

Although this calculation may look intimidating, it is very easy to perform with a pocket calculator and, of course, there is no need to write out the sum in full once you have Table A14.1 in front of you. The main effects of methionine and cystine are found in the same way.

To obtain the coefficients for the interactions, simply multiply together the coefficients in the corresponding main effects columns. Thus the LM column begins:

$$(-1)(-1) = +1; (+1)(-1) = -1; (-1)(+1) = -1 \text{ and so on.}$$

The LMC column of coefficients can be obtained by multiplying together the coefficients in the first *three* columns or, more conveniently, by multiplying the adjacent C and LM columns.

You will remember that the test for orthogonality given in Appendix 7 is that two contrasts are orthogonal if the products of their coefficients sum to zero. Since we have $\sum c = 0$ at the foot of the LM, LC, MC and LMC columns, these seven contrasts must form an orthogonal set, and when you have calculated the s.s. for each one (following the procedure given for lysine above) the seven components will add up to the diets s.s. (if not, you have made a mistake somewhere: try again!). This methodology is therefore *safer* (as well as quicker) than setting out two-way tables to find the interactions, since it is self-checking (if you calculate LM, LC and MC s.s. from two-way tables, you are then left to find LMC by difference, which means that it will include arithmetic mistakes made along the way).

When the sums are done, you should have a table like Table A14.2. The ANOVA tells us that both methionine and cystine have had highly significant effects on chick weight. Inspection of the means tells us that this effect is positive. There is a significant interaction between methionine and cystine and, again by examining the means, we can see that this is because their response is *non-additive*. When M and C are both added to the diet the response is no greater than when either one is added alone (and this statement is true whether or not lysine has been included in the diet).

Table A14.2. Preliminary subdivision of treatment s.s. for the data given in Exercise 7.1.

Source	d.f.	s.s.	m.s.	*F*
Treatments	7	1360.2581	194.3226	9.52***
L (= main effect of lysine)	1	93.8002	93.8002	4.59*
M	1	321.8852	321.8852	15.77***
C	1	321.8852	321.8852	15.77***
LM	1	80.8602	80.8602	3.96 $P<0.1$
LC	1	15.3002	15.3002	
MC	1	518.1102	518.1102	25.38***
LMC	1	8.4169	8.4169	
Error	35	714.4797	20.4137	
SEM = 1.845 g		SED = 2.609 g	LSD ($P = 0.05$) = 5.24 g	

The main effect of lysine is significant. Is this a favourable or an adverse effect of lysine on chick growth? Do not fall into the trap of looking only at the first two treatments (lysine versus control). The *main effect* of a factor means the response to that factor *averaged across* all other combinations of factors. If you look at the treatment means again you will see that lysine has apparently reduced growth rate when added by itself to the diet, but seems to have stimulated growth when added in conjunction with methionine or cystine or both. This is fairly typical behaviour for an amino acid that is *second limiting* in a diet. This accounts for the (nearly significant) LM interaction, though it should leave us wondering why the LC interaction is not also significant.

Reporting this experiment in terms of main effects and interactions will not be particularly helpful. Neither is a table festooned with superscripts indicating which individual treatment is significantly different (at $P = 0.05$, 0.01 or 0.001) from which other treatments. A further subdivision of the treatment s.s. is needed to clarify matters. This, in turn, requires a set of working hypotheses. The experimenter, having thought a lot about these data, wisely decides to draw a diagram representing his results. This is reproduced in Fig. A14.1.

The figure seems to suggest that:

1. Lysine, added alone to the basal diet, depresses growth.
2. Addition of a sulphur-containing amino acid (SAA), (i.e. either methionine or cystine or both) improves growth.
3. There is not much difference in growth rate between chicks on the M, C and MC diets.
4. Once the primary deficiency of SAA has been corrected, addition of lysine actually improves chick growth.
5. There is not much difference in growth between chicks on diets which contain both lysine and an SAA (i.e. diets LM, LC and LMC).

These five hypotheses can be tested. Questions 1, 2 and 4 are of the 'this versus that' type and will each involve 1 d.f. Questions 3 and 5 each require a comparison amongst 3 treatments and so each have 2 d.f. We could test the 'this versus that' propositions by means of Student's t, but a comparison amongst three means requires an F-test, so we might as well do all the testing within the ANOVA table. The appropriate s.s. are found as follows:

1. Lysine alone versus basal s.s. = $(909.7 - 920.3)^2/(2 \times 6) = \mathbf{9.3633}$.
2. Basal versus $\frac{1}{3}$(M + C + MC) s.s. =

$$\{3(920.3) - (970.2 + 979.0 + 960.1)\}^2/(12 \times 6) = \mathbf{305.8689}.$$

3. Amongst M, C and MC s.s. =

$$\{(970.2^2 + 979.0^2 + 960.1^2)/6\} - \{(970.2 + 979.0 + 960.1)^2/18\} = \mathbf{29.8144}.$$

4. (M + C + MC) versus (LM + LC + LMC) s.s. =

$$\{(970.2 + 979.0 + 960.1) - (1000.8 + 992.0 + 994.2)\}^2/(2 \times 18) = \mathbf{167.7025}.$$

5. Amongst LM, LC and LMC s.s. =

$$\{(1000.8^2 + 992.0^2 + 994.2^2)/6\} - \{(1000.8 + 992.0 + 994.2)^2/18\} = \mathbf{6.9911}.$$

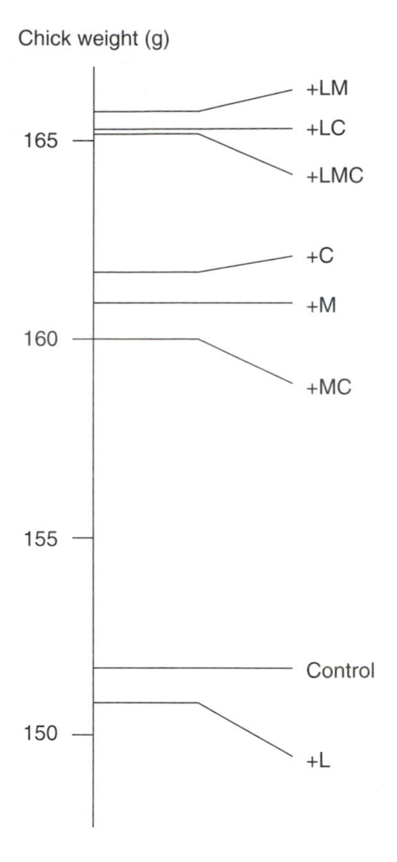

Fig. A14.1. An illustration of the treatment means derived from the data in Exercise 6.1.

If you add the s.s. which have been computed for these five effects, you will find that they do not add to the diet s.s. This is because the contrasts we have chosen, although logical, do not form an orthogonal set. Questions 1, 3, 4 and 5 are orthogonal but we need to substitute for question 2 the less interesting contrast ((basal + lysine alone) versus the rest) to complete an orthogonal set. The s.s. for this contrast is **1146.3867** and if you substitute this number for the 305.8689 s.s. obtained in answer to question 2 you will find that the s.s. do then add to the diet s.s. We can now set out the more informative ANOVA in Table A14.3 and conduct the necessary *F*-tests.

We can also confirm, for purposes of our report, that question 2 yields a m.s. which is highly significantly larger than error ($305.8689/20.4137 = 14.98^{**}$). We have now, in a sense, used 8 d.f. where only 7 d.f. were available, but as all the outcomes are clearly significant or clearly not significant there is no cause for concern that we may be drawing unreliable conclusions by asking too many questions.

Table A14.3. Final ANOVA of chick growth data from Exercise 7.1.

Source	d.f.	s.s.	m.s.	*F*
Replicates	5	225.3427	45.0685	2.21
Treatments	7	1360.2581	194.3226	9.52***
Control versus lysine alone	1	9.3633	9.3633	
(Control and control + Lys) versus rest	1	1146.3867	1146.3867	56.16***
(M,C and MC) versus (LM, LC and LMC)	1	167.7025	167.7025	8.18**
Amongst M, C and MC	2	29.8144	14.9072	
Amongst LM, LC and LMC	2	6.9911	3.4955	
Error	35	714.4797	20.4137	

You might wonder about the comparatively small m.s. for the last contrast in Table A14.3. Are the LM, LC and LMC means suspiciously close together? We can compare this variance with the error m.s. (remembering to divide the *greater* m.s. by the *lesser*) to obtain an F-ratio of 5.84 with 35 and 2 d.f. As the lesser m.s. has only 2 d.f. the 5 percentage point for F is 19.5 and there is no cause for alarm. With so few d.f. for the divisor, a treatment m.s. (or a component) can easily be as small as one-sixth of the error term. The odds against it are only about 5:1.

A final report on this trial might read: 'Chicks fed a cereal-based diet in which the only protein supplement was noog meal grew at an average of 11.5 g day^{-1} from 7 to 21 days. Addition of lysine to this diet did not significantly affect growth rate. Addition of a sulphur amino acid (either methionine or cystine) resulted in a 6% improvement in growth rate (highly significant). If methionine and cystine were both added there was no further improvement in growth. Supplementation with both lysine and a sulphur-containing amino acid resulted in a further improvement of 2.6% in growth rate above that obtained with sulphur-containing amino acid addition alone. Since chicks can derive cystine from methionine but not vice versa, the basal diet must have been first-limiting in cystine (if it were deficient in methionine one would expect no response to added cystine). The second limiting amino acid in the diet was lysine.'

Appendix 15

Bartlett's Test

We will illustrate the procedures for testing whether treatment variances are homogeneous by using the data illustrated in Fig. 8.1. The variances for the five treatments are listed in Table A15.1.

Table A15.1. Variances of plasma LH (ng ml^{-1}) from an experiment in which five doses of GnRH were applied. Twelve animals were used for each of the five treatments ($t = 5$, $n = 12$).

Dose (coded)	Variance = m.s. = s^2	ln s^2
0	1.46	0.37844
1	3.47	1.24415
2	4.92	1.59331
3	9.15	2.21375
4	8.59	2.15060
	$\Sigma s^2 = 27.59$	$\Sigma \ln s^2 = 7.58025$
	$\bar{s}^2 = 5.518$	
	$\ln \bar{s}^2 = 1.70802$	

Step 1. Test the extremes

The ratio of the largest variance to the smallest one is an F-ratio which can be checked in the tables to see whether those two variances are significantly different.

$F = 9.15/1.46 = 6.27$ (11 and 11 d.f.), $P < 0.01$.

In this case the two extremes differ more than would be expected if they were two sample variances drawn at random from a normal population of variances. But, since there are five variances in our sample, this test is not conclusive. We need to ask whether the set of five differ more than would be expected by normal sampling variation. If each variance is based on the same sample size, proceed to **Step 2**. If the sample sizes differ, proceed to **Step 3**.

Step 2. Testing homogeneity of variance where the variances come from samples of equal size

Bartlett's test involves computing the difference between (t times the natural logarithm of the average variance) and (the sum of the natural logarithms of each of the variances). This quantity, multiplied by the d.f. attached to each variance, follows a χ^2 distribution (approximately). The first steps of the calculation are set out in Table A15.1.

$t.\ln \bar{s}^2 = 5(1.70802) = 8.54010$

$\Sigma \ln s^2 = 7.58025$

difference $= 0.95985$.

Approximate $\chi^2 = (n - 1)(t.\ln \bar{s}^2 - \Sigma \ln s^2)$

$= (11)(0.95985) = 10.558_{(\text{d.f.} = t - 1 = 4)}$, $P < 0.05$.

Thus the five variances differ to a significantly greater extent than would be expected by random sampling.

Unfortunately, this is not quite the end of the matter. If the variances are based on very large sample sizes, the probability estimate is near enough; but for limited sample sizes it is necessary to apply a correction factor if the approximate result is close to a probability boundary.

Correction factor, $C = 1 + [(t + 1)/\{3t(n - 1)\}] = 1 + (6/165) = 1.03636$.

Corrected χ^2 = approximate $\chi^2/C = 10.558/1.03636 = 10.188_{(4\ \text{d.f.})}$, $P < 0.05$.

If the approximate χ^2 is more than 10% above a probability boundary, there is no need to bother with the correction, since even for very low values, such as $t = 3$ and $n = 4$, $1/C = 0.9$.

Step 3. Testing homogeneity of variance where the variances come from samples of unequal size

Bartlett's test still applies, but now the variances have to be weighted by their respective d.f. Table A15.2 gives the necessary data and the initial steps in the calculations.

$\bar{s}^2 = \Sigma\text{s.s.}/\Sigma(n - 1) = 270.10/50 = 5.402$ (weighted mean s^2).

$\ln \bar{s}^2 = 1.68677$.

$\Sigma(n - 1).\ln \bar{s}^2 = 50(1.68677) = 84.33850$.

$\Sigma\{(n - 1)\ln s^2\} = 73.95246$.

Approx. χ^2 = difference $= 10.386_{(\text{d.f.} = t - 1 = 4)}$, $P < 0.05$.

Correction factor $= 1 + [1/(3)(t - 1)][\{\Sigma[1/(n - 1)]\} - \{1/\Sigma(n - 1)\}]$

$= 1 + [1/(3)(4)][0.51912 - (1/50)]$

$$= 1.04159.$$

$$\text{Corrected } \chi^2 = 10.386/1.04159 = 9.971_{(4\ \text{d.f.})},\ P < 0.05.$$

Table A15.2. Sums of squares and mean squares from an experiment in which five doses of GnRH were tested, with differing numbers of animals (n) yielding results for each treatment ($t = 5$, n is variable).

Dose (coded)	s.s.	d.f. $= n - 1$	$1/(n - 1)$	m.s. $= s^2$	$\ln s^2$	$(n - 1)\ln s^2$
0	16.06	11	0.09091	1.46	0.37844	4.16284
1	41.64	12	0.08333	3.47	1.24415	14.92980
2	34.44	7	0.14286	4.92	1.59331	11.15317
3	100.65	11	0.09091	9.15	2.21375	24.35125
4	77.31	9	0.11111	8.59	2.15060	19.35540
	270.10	50	0.51912			73.95246
	Σs.s.	$\Sigma(n - 1)$	$\Sigma 1/(n - 1)$			$\Sigma(n - 1)\ln s^2$

Appendix 16

Answers to Exercise 8.1

The data from Table 8.3 are reproduced in Table A16.1, which also shows the variances calculated separately for each treatment.

Table A16.1. Plasma lactic acid concentrations adjusted to remove block effects.

	Diet		
Replicate	Hay	Grass	Maize
1	0.2533	0.5833	0.8633
2	0.3933	0.3033	1.0033
3	0.1433	0.3133	1.2433
4	0.3267	0.4767	0.8967
5	0.1600	0.1400	1.4000
6	0.1933	0.3433	1.1633
Totals	1.4699	2.1599	6.5699
s^2	0.0097955	0.0235113	0.0442199

Step 1. Compare the extremes

$$F = 0.0442199/0.0097955 = 4.51_{(5,5\text{ d.f.})},\ 0.10 > P > 0.05.$$

The difference between the largest and the smallest variance is not significant but is suspiciously large.

Step 2. Run Bartlett's test

Table A16.2 lists the opening calculations.

$$\text{Approximate } \chi^2 = (n-1)(t.\ln \bar{s}^2 - \Sigma \ln s^2) = 5(0.527451) = 2.637 \text{ (2 d.f.)},\ 0.30 > P > 0.20.$$

Variances as diverse as these would occur about once in four trials and so our suspicion of heterogeneity was not justified.

Table A16.2. Bartlett's test applied to variances in Table A16.1.

Treatment	s^2	$\ln s^2$
Hay	0.0097955	−4.625832
Grass silage	0.0235113	−3.750274
Maize silage	0.0442199	−3.118580
	Σs^2 = 0.0775267	$\Sigma \ln s^2$ = −11.494686
	$\bar{s}^2$ = 0.0258422	
	$\ln \bar{s}^2$ = −3.655745	$t.\ln \bar{s}^2$ = −10.967235
		Difference = 0.527451

Appendix 17

Answers to Exercise 9.1

Answer to question 1

You were told that 64 cows were allocated to eight treatments in four randomized complete blocks and that a covariance on initial milk fat values was carried out. This should enable you to reconstruct the ANOVA framework, as in Table A17.1, confirming the 52 d.f. for error shown in Table 9.1.

Table A17.1. The ANOVA reconstructed as far as it is possible to go with the data in Table 9.1.

Source	d.f.		s.s.		m.s.	*F*
Initial yield	1					
Blocks	3					
Treatments	7		1749.11		249.873	6.71^{***}
Hay (*ad lib.* versus restricted)		1		216.09	216.09	5.79^{*}
Concentrates		3		1512.19	504.063	13.50^{***}
Interaction		3		20.83	6.943	
Error	52		1940.89		37.3248	
Total	63					

Given that SEM = 2.16 with eight cows on each treatment, we can calculate:

$$s = 2.16(\sqrt{8}) = 6.1094;\ s^2 = \text{error m.s.} = 37.3248.$$

The treatments s.s. can be obtained from the treatment means as follows:[a]

$$\text{treatment s.s.} = \Sigma(\text{treatment totals})^2/8 - (\text{Grand total})^2/64$$

$$= \Sigma\{8(\text{treatment mean})\}^2/8 - \{8(\Sigma\text{treatment means})\}^2/64$$

$$= 8\Sigma(\text{treatment means})^2 - (\Sigma\text{treatment means})^2$$

$$= 8(25.6^2 + 32.0^2 \ldots + 40.8^2) - (25.6 + 32.0 \ldots + 40.8)^2 = 1749.11.$$

[a] More generally, for a series of t treatment means, each based on n replicates, the treatment s.s. = $n\Sigma(\text{treatment means})^2 - (n/t)[\Sigma(\text{treatment means})]^2$.

The s.s. for Hay (*ad lib.* versus restricted), concentrates and the interaction can be found by analogous methods from the treatment means in Table 9.1. This gives you the ANOVA in Table A17.1.

The effect of feeding hay *ad libitum* or in restricted amounts was significant and the effect of concentrate formulation on milk fat was highly significant. There was no interaction between these two factors.

Answers to question 2

After using a standard computer package to fit the models specified, you will need to multiply the resulting s.s. by $n = 8$ to obtain values consonant with the error term in Table A17.1. This should give you the results in Table A17.2.

Table A17.2. Regression models fitted to the milk fat data in Table 9.1.

Source	d.f.	s.s.	m.s.	*F*
Treatments	**7**	**1749.11**	**249.873**	**6.71*****
Model 1				
Single linear regression	1	1383.300	1383.300	37.06***
Residual	6	365.810	60.968	1.63 NS
Model 2				
Two parallel lines	2	1595.634	797.817	21.37***
Residual	5	153.475	30.695	–
Difference: model 2 versus 1	1	212.334	212.334	5.60*
Model 3				
One quadratic curve	2	1455.628	727.814	19.50***
Residual	5	293.482	58.696	1.57 NS
Difference: model 3 versus 1	1	72.328	72.328	1.94 NS
Model 4				
Parallel curves	3	1634.540	544.847	14.60***
Residual	4	114.570	28.643	–
Difference: model 4 versus 3	1	178.912	178.912	4.79*
Difference: model 4 versus 2	1	38.905	38.905	1.04 NS
Error	**52**	**1940.89**	**37.3248**	

The single linear regression accounts for 79% of the variation amongst the treatments, but a pair of parallel lines is a significantly better fit. Note that significance at $P = 0.05$ is not crucial in this context. Even if the odds were only 5:1 in favour of the more complex model, you might well prefer it. With other data you might want to test a model fitting two independent straight lines (i.e. with differing slopes for hay *ad lib.* and restricted), but for the present data that would fairly obviously be a waste of time.

A single curve through all the treatment means gives a good fit, but not as good as a pair of parallel curves (the difference between models 3 and 4 is significant). That leaves us to choose between a pair of straight lines and a pair of curves. Since the difference between model 4 and model 2 is no greater than the error term, Ockham's razor[a] comes into operation and we choose two parallel straight lines as the best model to represent our results.

Note that there is nothing inherently unlikely about linear responses in this case. The range of inputs is limited absolutely by starch concentrations of 0 and 1000 g kg^{-1} and, using practical (as opposed to purified) ingredients and allowing that the diet must contain some protein, values of 140 and 430 g starch kg^{-1} are about as extreme as can be investigated. There is no a priori reason to expect a diminishing returns curve over this range of inputs.

Answer to question 3

There are various possible answers, but one fairly obvious step would be to calculate the proportion of starch in each complete diet, taking the composition and intake of hay into account. You would expect to find that the (presumably) higher intakes of hay under *ad libitum* hay feeding resulted in lower proportions of starch in the complete diet (for any one concentrate feed) and this might account quantitatively for the separation of the two response lines. There are two ways of defining a new variable: (1) starch as a proportion of total diet DM and (2) ME from starch as a proportion of total dietary ME. Where the hay: concentrate ratio is free to vary, these two variables yield slightly different relationships and you might find that one or the other gave the best resolution of the eight data points for milk fat into a single linear regression.

[a] William of Ockham wrote (in Latin in about 1320 AD) that 'entities are not to be multiplied beyond necessity', meaning that, when faced with two or more hypotheses which equally well explain the facts, the simplest explanation is to be preferred.

Appendix 18

Example of Analysis of Covariance

The data from Table 10.1 will be used to illustrate how an analysis of covariance works. The numbers are perfectly manageable on a pocket calculator, but covariance analysis is much more conveniently done with a computer package, especially where there is a complex design structure. If there are blocks or a Latin square or split plots, these must be taken into account in the analysis of covariance. This example is simplified by assuming that there was no blocking in the experiment to be analysed. The data from Table 10.1 are reproduced here as Table A18.1 for convenience.

Table A18.1. Preliminary yields (x, kg day^{-1}), measured in 2 weeks before treatments were applied, and yields (y, kg day^{-1}) during 36 weeks of treatment for 16 dairy cows allocated to four dietary treatments.

Treatment		Individual yields				Totals	Means
A	x	26.1	20.4	17.5	23.4	87.4	21.85
	y	21.5	17.9	17.6	25.7	82.7	20.68
B	x	21.4	25.6	23.0	16.9	86.9	21.73
	y	18.4	25.1	19.8	15.8	79.1	19.78
C	x	16.4	24.8	19.0	18.5	78.7	19.68
	y	12.2	21.6	18.3	19.1	71.2	17.80
D	x	26.2	16.9	18.7	21.3	83.1	20.78
	y	21.9	12.0	15.1	18.0	67.0	16.75
				Grand totals	x	336.1	
					y	300.0	

The ANOVA structure will be as shown in Table A18.2. The sums of squares for the x and y variables are calculated in the usual way (see Appendix 3 if in doubt).

The sums of products (s.p.) are calculated in a similar way, but multiplying x by y instead of squaring each one. Thus:

Table A18.2. The analysis of variance and covariance table.

Source	d.f.	s.s. x	s.p. xy	s.s. y	m.s. y
Treatments	3	12.2169	16.3525	38.635	12.878
Error	12	169.3725	155.3625	190.245	15.854
Total	15	181.5894	171.7150	228.880	

Correction factor = $\{(\sum x)(\sum y)\}/N = \{(336.1)(300.0)\}/16 = 6301.875$.

Treatment s.p. = $\dfrac{(87.4)(82.7) + \ldots (83.1)(67.0)}{4} - \text{CF} = \mathbf{6.3525}$.

Total s.p. = $(26.1)(21.5) + (20.4)(17.9) + \ldots (21.3)(18.0) - \text{CF} = \mathbf{171.7150}$.

Error s.p. by difference: $171.7150 - 16.3525 = \mathbf{155.3625}$.

Calculate the correlation coefficient, r.

$r = (\text{error s.p. } xy)/\sqrt{(\text{error s.s. } x)(\text{error s.s. } y)}$

$= 155.3625/\sqrt{(169.3725)(190.245)} = \mathbf{+0.8655}^{***}$.

The significance of a correlation coefficient can be looked up in statistical tables or you can use the F tables in Appendix 25 taking the column for 1 and error d.f. (12 in this case) after calculating:

$F = \{(\text{error d.f.})r^2\}/(1 - r^2) = 35.8\ (P < 0.001)$.

Calculate the regression coefficient, b.

$b = (\text{error s.p. } xy)/(\text{error s.s. } x)$

$= 155.3625/169.3725 = \mathbf{+0.9173}$ kg y/kg x.

Calculate the residual sums of squares as follows:

Total s.s.$(y - bx)$ = total s.s. y − (total s.p. $xy)^2$/total s.s. x

$= 228.88 - (171.715)^2/181.5894 = \mathbf{66.502}$.

Error s.s. $(y - bx)$ = Error s.s. y − (error s.p. $xy)^2$/error s.s. x

$= 190.245 - (155.3625)^2/169.3725 = \mathbf{47.734}$.

Rather surprisingly, treatment s.s.$(y - bx)$ is calculated by difference, not from adjusted treatment totals or means as you might have suspected:

Treatment s.s. $(y - bx)$ = total s.s. $(y - bx)$ − error s.s. $(y - bx)$.

These calculations lead to a table of residual variance (Table A18.3) in which the degrees of freedom for total and error are reduced by one, corresponding to the 1 d.f. associated with the calculation of a linear regression coefficient.

Table A18.3. Analysis of residual variance.

Source	d.f.	s.s.($y - bx$)	m.s.	F
Treatments	3	18.768	6.256	1.44 NS
Error	11	47.734	4.339	
Total	14	66.502		

Unfortunately, the use of the error mean square from Table A18.3 to compare adjusted treatment means is not straightforward because the difference between two adjusted means is

$$y_1 - y_2 - b(x_1 - x_2)$$

and the variance of this quantity includes the variance of b as well as the square of the quantity $(x_1 - x_2)$. For any particular pair of adjusted treatment means the SED is:

$$\sqrt{\text{residual error m.s. } \{2/n + (x_1 - x_2)^2/\text{error s.s. } x\}}.$$

As a compromise it is usual to calculate an effective residual mean square:

$$\text{effective residual m.s.} = \text{residual error m.s.} \left(\frac{1 + \text{treatment m.s. } x}{\text{error s.s. } x}\right)$$

$$= 4.339\left(1 + \frac{4.0723}{169.3725}\right) = 4.443$$

The approximate SED of two adjusted treatment means is then

$$\sqrt{\frac{(2)(4.443)}{4}} = 1.490 \text{ kg.}$$

If the result of testing with this compromise SED comes close to a probability boundary, the test should be repeated using the exact SED for the pair of treatments in question. The adjusted treatment means are given in Table A18.4.

Table A18.4. Mean yields and adjusted treatment means (kg day^{-1}).

Treatment	Mean x	Mean y	Adjusted treatment means ($y - b(x - \bar{x})$)
A	21.85	20.68	19.91
B	21.73	19.78	19.12
C	19.68	17.80	19.02
D	20.78	16.75	16.96
	$\bar{x} = 21.01$		$b = 0.9173$
			Approximate SEM ± 1.05

Appendix 19

Answers to Exercise 10.1

Table A19.1 gives an analysis of the data for liveweight gain from this exercise. It is clear that pigs fed the jack beans grew more slowly than the controls and the depression in growth rate was linearly related to the dose of jack beans included in the diet.

Table A19.1. ANOVA of data for liveweight gain (g day^{-1}) of pigs as listed in Table 10.5.

Source	d.f.		s.s.		m.s.	*F*
Diets	2		164,305.33		82,152.67	5.82^{**}
Linear regression		1		164,025.00	164,025.00	11.62^{**}
Deviations		1		280.33	280.33	–
Error	21		296,474.00		14,117.809	
Total	23		460,779.33			

The corresponding analysis for feed intake is in Table A19.2. Feed consumption was also depressed by inclusion of jack beans in the diet and, again, the response was linearly related to dose.

The analysis of covariance is presented in Table A19.3. After adjustment of growth rate for differences in feed intake, there are still differences amongst

Table A19.2. ANOVA of data for feed intake (g day^{-1}) of pigs as listed in Table 10.5.

Source	d.f.		s.s.		m.s	*F*
Diets	2		582,948.58		291,474.29	3.30 $P<0.10$
Linear regression		1		582,932.25	582,932.25	6.61^{*}
Deviations		1		16.33	16.33	–
Error	21		1,852,605.22		88,219.29	
Total	23		2,435,553.80			

Table A19.3. Analysis of covariance of feed intake and growth rate for the pig data from Table 10.5.

Source	d.f.	s.p. *xy*	s.s. $y - bx$	d.f.	m.s.	*F*
Diets	2	309,285.17	20,281.95	2	10,140.97	1.85 NS
Regression	1		20,049.30	1	20,049.30	3.67 ($P<0.07$)
Deviations	1		232.65	1	232.65	–
Error	21	588,778.50	109,353.66	20	5,467.68	
Total	23	898,063.67	129,635.61			
		$b = 0.3178$ g g^{-1}; $r^2 = 0.631$; $r = 0.794^{***}$				

diets which, though not significant at $P = 0.05$ by *F*-test, still indicate a linear regression ($P \approx 0.07$). The reduction in growth rate associated with 160 g kg^{-1} jack beans was 18% before adjustment, but is only 8% at constant feed intake. We may infer that reduced feed intake is a major reason for the poor performance on the diets containing jack beans, but does not account for all of it. Adjusted treatment means are presented in Table A19.4.

It is never possible to conclude with certainty from an analysis of covariance whether growth rate is being driven by feed intake, or whether differences in feed intake are largely the consequence of different growth rates. However, in the present case, feed intakes of individual pigs on different treatments overlap and we are therefore not required to make a dangerous extrapolation. It is an *observation*, not a conjecture, that those pigs on the jack bean diets having feed intakes in the range 2400–2800 g day^{-1} achieved lower mean growth rates than those on the control diet with feed intakes in the same range. This is a powerful indicator that some of the toxic effect of the jack beans is due to a metabolic effect independent of feed intake; but the observed depression in growth (which is 18% overall, but only 8% at the same feed intake) could be mainly due to an aversion to eating the beans and only to a lesser extent be a result of the later adverse metabolic consequences of taking the jack bean toxins into the system. Aversion may not be just a question of unpleasant experiences at the palate (taste or smell) but may reflect early reactions from the gut to the intake of toxic compounds.

Table A19.4. Treatment means for growth rate adjusted for differences in mean feed intake (kg day^{-1}).

Jack beans in diet (g kg^{-1})	Mean feed intake (kg day^{-1})	Mean growth rate (kg day^{-1})	Growth rate adjusted to mean feed intake (kg day^{-1})
0	2.848	1.122	1.062
80	2.659	1.028	1.028
160	2.467	0.919	0.979
	Mean 2.658; approximate SEM: 0.0281		

Two experimental procedures can help to elucidate these problems. The first, *pair feeding*, has been widely used in animal nutrition. It involves matching pairs of animals at the outset, allocating them to control and treatment, and then giving the control animal as much feed as its treated pair consumes on the preceding day. With replicated pairs, this should result in identical mean feed intakes for control and treated animals. Their growth rates can then be directly compared under conditions where the feed supply is the same. The practical difficulties of managing this procedure are, however, considerable. Suppose the mean difference in feed intake between control and treated animals under *ad libitum* conditions is 10%. It would not then be surprising if some of the control animals receiving rationed quantities of feed failed to consume all that was given to them. Add to that the fact that there will be considerable day-to-day variation in the *ad libitum* intake of the treated animal and you have a whole series of headaches in managing the experiment in such a way that paired animals do end up having received the same amount of feed. It is almost inevitable that some control animals will clear their trough in a few hours while others leave stale feed accumulating in the feeder. Pair feeding is fine in principle, but very difficult and time consuming to manage in practice.

A second type of experiment which conveys a good deal of useful information in a short time is the *immediate feed intake response trial*. By offering naïve animals the control or treated feed and recording at frequent intervals (half-hourly or hourly for rats or chicks, but every 5 or 10 minutes for pigs accustomed to meal feeding) how much they consume in the first 24 hours, you can get a very good idea of whether the primary response is a depression of intake. You can also measure the magnitude of that initial depression. With astringent or toxic materials, the initial reduction is often greater than the depression observed over a period of weeks. The animals do not like the feed, but they have to eat something! You may also be able to judge whether the problem is at the palate or follows after the absorption of undesirable compounds. With toxins at low doses, the animal may eat normally for the first hour but then begin to show an adverse reaction. If the two groups of animals eat the same amount of feed in the first 24 hours, you can reasonably conclude that the reduction in feed intake observed in a longer trial is a secondary consequence of the reduced growth rate which that diet supports and not primarily the result of an aversion to the feed.

Appendix 20

Answers to Exercise 11.1

Because there is no blocking in the experiment described in Table 11.2 we can do a simple one-way ANOVA.

The treatment s.s. will be:

$(617^2/8) + (380^2/10) + (-1124^2/5) - CF = 314{,}000.06.$

Control versus supplements s.s. $= (-1124^2/5) + (997^2/18) - CF = 307{,}196.66.$

Between supplements s.s. $= (617^2/8) + (380^2/10) - (997^2/18) = 6803.40.$

Total s.s. by the usual method. Error by difference.

The ANOVA is shown in Table A20.1.

Table A20.1. ANOVA for the data in Table 11.2.

Source	d.f.		s.s.	m.s.	*F*
Treatments	2		314,000.06	157,000	33.21*
Control versus supplements		1	307,196.66	307,197	64.98***
Between supplements		1	6,803.40	6,803	1.44 NS
Error	20		94,549.68	4,727.48	
Total	22		408,549.74		

The subdivision of the treatment s.s. shown in Table A20.1 answers all the questions. There was a highly significant improvement in weight gain (i.e. avoidance of weight loss) due to supplementary feeding through the dry season but no difference between the two supplements tested.

However, for publication, you may be required to produce SEDs to go with your table of means, in which case you need to calculate:

SED supplement 1 versus 2 $= \sqrt{\{(1/8) + (1/10)\}.\{4727.48\}} = 32.61$ g day^{-1}

SED supplement 1 versus control $= \sqrt{\{(1/8) + (1/5)\}.\{4727.48\}}$
$= 39.20$ g day^{-1}

$$\text{SED supplement 2 versus control} = \sqrt{\{(1/10) + (1/5)\}.\{4727.48\}}$$
$$= 37.66 \text{ g day}^{-1}$$

A suitable summary of these results is given in Table A20.2.

Table A20.2. Results of an experiment comparing two supplements fed to beef cows on range throughout the dry season.

	Treatment		
	A Control	B Supplement 1	C Supplement 2
No. of cows	5	8	10
Mean liveweight change (g day^{-1})	−224.8	+77.1	+38.0
SEM	83.3	24.3	21.7
	Difference (g day^{-1})	SED	*P*
A versus B	301.9	39.2	< 0.001
A versus C	262.8	37.7	< 0.001
B versus C	39.1	32.6	> 0.2

Appendix 21

Answers to Exercise 11.2

The ANOVA of the data from Table 11.3, taking account of blocks, is given in Table A21.1.

Table A21.1. ANOVA for the data in Table 11.3.

Source	d.f.	s.s.	m.s.	*F*
Blocks	9	54,631.49	6,070.17	1.67 NS
Treatments	2	216,148.48	108,074.24	29.78^{***}
Error	11	39,918.18	3,628.93	
Total	22	408,549.74		

Notice that, with least squares estimates of s.s. in an unbalanced design, the component s.s. do *not* add up to equal the total s.s.. Although the blocks m.s. is not significantly larger than error, removal of the block effects in this example has reduced the error m.s. to 77% of what it was in Exercise 11.1, where the blocking was ignored. This is a case where a computer package capable of carrying out a least squares analysis of unbalanced data proves to be useful, although it does not change the conclusions about the treatments.

Appendix 22

Answers to Exercise 12.1

1. It is unlikely that heifer 19 made a true gain of 27 kg in the 14 days before this weighing, but number 22 also apparently made a large gain in weeks 3 and 4 and then lost weight in weeks 5 and 6. You could equally well say that the week 6 weight of number 19 was aberrant. There is no basis for rejecting the recorded 440 kg as a mistake.

Note that, when you estimate weight gain by regression, these ups and downs do not make much difference to the long-term trend: but we should have much less confidence that 440 kg (following 413 kg) was a true record for number 19 if it was not followed by further weighings. This emphasizes the hazard of relying on final weights to estimate rate of gain.

2. The analysis is false because the error based on 126 d.f. is not valid for testing treatment differences. These are sequential measurements on the same individuals. There can only be 21 d.f. for comparing differences between three sets of eight animals.

3. You can use a computer program to estimate the regressions, or you can calculate b, the regression slope for each animal quite easily, using orthogonal polynomials for six *equally spaced* treatments (see Appendix 7). Thus, rate of gain for heifer number 1 is:

$$b = \{-5(368) - 3(380) - 1(377) + 1(389) + 3(413) + 5(426)\}/70$$

$$= 5.729 \text{ kg week}^{-1}.$$

The full results (expressed as gain per day) are set out in Table A22.1 and the ANOVA is in Table A22.2.

'Pens' do not appear in this analysis because pens are not orthogonal with treatments. If you have access to a suitable computer package for analysing non-orthogonal designs, you can estimate a component for pens, but it will make little difference to the result with these data.

Although, if you draw a graph of weight against time, treatment 3 appears to be lagging behind the other two treatments throughout the trial; this only reflects random variation amongst the animals. If you had conducted the *false* analysis set out in question 2 of this exercise, you would have concluded that

Table A22.1. Rate of liveweight gain (g day^{-1}) for the animals listed in Table 12.3, estimated by linear regression.

Treatment					
1		2		3	
Animal	Gain	Animal	Gain	Animal	Gain
3	1100	2	1086	1	818
6	947	4	776	7	655
8	978	5	1200	9	1190
10	1069	11	1133	12	1192
15	1282	18	986	13	904
16	1124	19	1149	14	1022
20	1108	22	1143	17	943
23	1047	24	1092	21	1276

treatment differences were significant. This is because the inferior growth rate on treatment 3 is consistent in time. However, the true analysis in Table A22.2 tells us that the apparent inferiority of treatment 3 is due to random allocation of heifers with slightly poorer growth potential to that treatment at the outset.

Table A22.2. ANOVA of rate of liveweight gain (g day^{-1}) estimated by linear regression.

Source	d.f.	s.s.	m.s.	*F*
Treatments	2	31,514.58	15,757.292	–
Error	21	514,104.75	24,481.178	
Total	23	545,619.33		

4. The CV for weight gain from Table A22.2 is 14.7%. If you conduct a similar ANOVA of weight gain estimated by difference using initial and final weights, you will find a CV of 15.1%. In this case there is very little difference in the precision of the two methods. However, regression analysis is, generally, a much safer procedure (provided that liveweight gain is approximately linear throughout the trial) and it often does give a worthwhile improvement in precision.

Appendix 23

Example of Matrix Used to Fit Constants for Trials

Data: 15 trials (illustrated in Fig. 14.3) published between 1951 and 1996 showing the effect of photoperiod, held constant from hatching until after sexual maturity, on mean age at first egg in chickens.

Model: $y = b_0 + b_1x + b_2x^2 + k_i$

where y = age at first egg (days); x = photoperiod (h), and k_i = a constant for the ith trial; $i = 1 \ldots 14$.

Matrix for regression analysis

Trial	y	x	x^2	k_1	k_2	k_3	k_4	k_5	k_6	k_7	k_8	k_9	k_{10}	k_{11}	k_{12}	k_{13}	k_{14}
1	160.0	6	36	0	0	0	0	0	0	0	0	0	0	0	0	0	0
	163.2	14	196	0	0	0	0	0	0	0	0	0	0	0	0	0	0
	164.1	22	484	0	0	0	0	0	0	0	0	0	0	0	0	0	0
2	158.3	6	36	1	0	0	0	0	0	0	0	0	0	0	0	0	0
	147.9	14	196	1	0	0	0	0	0	0	0	0	0	0	0	0	0
	150.6	22	484	1	0	0	0	0	0	0	0	0	0	0	0	0	0
3	166.0	6	36	0	1	0	0	0	0	0	0	0	0	0	0	0	0
	154.5	10	100	0	1	0	0	0	0	0	0	0	0	0	0	0	0
4	166.2	6	36	0	0	1	0	0	0	0	0	0	0	0	0	0	0
	153.6	10	100	0	0	1	0	0	0	0	0	0	0	0	0	0	0
	157.4	14	196	0	0	1	0	0	0	0	0	0	0	0	0	0	0
	162.6	18	324	0	0	1	0	0	0	0	0	0	0	0	0	0	0
5	150.7	8	64	0	0	0	1	0	0	0	0	0	0	0	0	0	0
	147.4	11	121	0	0	0	1	0	0	0	0	0	0	0	0	0	0
6	141.1	8	64	0	0	0	0	1	0	0	0	0	0	0	0	0	0
	134.0	10	100	0	0	0	0	1	0	0	0	0	0	0	0	0	0
	137.3	13	169	0	0	0	0	1	0	0	0	0	0	0	0	0	0
	134.6	18	324	0	0	0	0	1	0	0	0	0	0	0	0	0	0
7	146.1	8	64	0	0	0	0	0	1	0	0	0	0	0	0	0	0
	137.3	10	100	0	0	0	0	0	1	0	0	0	0	0	0	0	0
	145.6	13	169	0	0	0	0	0	1	0	0	0	0	0	0	0	0
	145.7	18	324	0	0	0	0	0	1	0	0	0	0	0	0	0	0
8	158	0	0	0	0	0	0	0	0	1	0	0	0	0	0	0	0
	157	2	4	0	0	0	0	0	0	1	0	0	0	0	0	0	0
	159	4	16	0	0	0	0	0	0	1	0	0	0	0	0	0	0
	153	6	36	0	0	0	0	0	0	1	0	0	0	0	0	0	0
	151	8	64	0	0	0	0	0	0	1	0	0	0	0	0	0	0
	151	10	100	0	0	0	0	0	0	1	0	0	0	0	0	0	0
	150	12	144	0	0	0	0	0	0	1	0	0	0	0	0	0	0
	147	14	196	0	0	0	0	0	0	1	0	0	0	0	0	0	0
9	175	0	0	0	0	0	0	0	0	0	1	0	0	0	0	0	0
	168	2	4	0	0	0	0	0	0	0	1	0	0	0	0	0	0
	154	4	16	0	0	0	0	0	0	0	1	0	0	0	0	0	0
	156	6	36	0	0	0	0	0	0	0	1	0	0	0	0	0	0
	152	8	64	0	0	0	0	0	0	0	1	0	0	0	0	0	0
	148	12	144	0	0	0	0	0	0	0	1	0	0	0	0	0	0
10	169.0	0	0	0	0	0	0	0	0	0	0	1	0	0	0	0	0
	151.0	6	36	0	0	0	0	0	0	0	0	1	0	0	0	0	0
11	155	1	1	0	0	0	0	0	0	0	0	0	1	0	0	0	0
	148	3	9	0	0	0	0	0	0	0	0	0	1	0	0	0	0
	145	9	81	0	0	0	0	0	0	0	0	0	1	0	0	0	0
	140	13	169	0	0	0	0	0	0	0	0	0	1	0	0	0	0
12	184.0	6	36	0	0	0	0	0	0	0	0	0	0	1	0	0	0
	173.0	14	196	0	0	0	0	0	0	0	0	0	0	1	0	0	0
13	167.0	6	36	0	0	0	0	0	0	0	0	0	0	0	1	0	0
	169.5	14	196	0	0	0	0	0	0	0	0	0	0	0	1	0	0
14	170.5	14	196	0	0	0	0	0	0	0	0	0	0	0	0	1	0
	171.5	24	576	0	0	0	0	0	0	0	0	0	0	0	0	1	0
15	166.0	14	196	0	0	0	0	0	0	0	0	0	0	0	0	0	1
	164.0	24	576	0	0	0	0	0	0	0	0	0	0	0	0	0	1

Appendix 24[a]

Table of χ^2

A χ^2 value with a specified number of degrees of freedom is significant at the 5-percentage point ($P = 0.05$) if it is *greater than* the number tabulated in the column headed $P = 5$.

Degrees of freedom	*P*					
	50	10	5	2.5	1	0.1
1	0.45	2.71	3.84	5.02	6.64	10.8
2	1.39	4.61	5.99	7.38	9.21	13.8
3	2.37	6.25	7.82	9.35	11.3	16.3
4	3.36	7.78	9.49	11.1	13.3	18.5
5	4.35	9.24	11.1	12.8	15.1	20.5
6	5.35	10.6	12.6	14.5	16.8	22.5
7	6.35	12.0	14.1	16.0	18.5	24.3
8	7.34	13.4	15.5	17.5	20.1	26.1
9	8.34	14.7	16.9	19.0	21.7	27.9
10	9.34	16.0	18.3	20.5	23.2	29.6
12	11.3	18.5	21.0	23.3	26.2	32.9
15	14.3	22.3	25.0	27.5	30.6	37.7
20	19.3	28.4	31.4	34.2	37.6	45.3
24	23.3	33.2	36.4	39.4	43.0	51.2
30	29.3	40.3	43.8	47.0	50.9	59.7
40	39.3	51.8	55.8	59.3	63.7	73.4
60	59.3	74.4	79.12	83.3	88.4	99.6

[a] Appendices 24, 25 and 26 are reproduced from Mead, Curnow and Hasted's *Statistical Methods in Experimental Biology and Agriculture*, published by Chapman & Hall, London, with kind permission from Kluwer Academic Publishers.

Appendix 25

F-ratio Tables

F values for the 5-percentage point ($P = 0.05$)

n_1 = degrees of freedom for the larger of two mean squares;
n_2 = degrees of freedom for the smaller mean square.

	n_1										
n_2	1	2	3	4	5	6	7	8	10	12	24
2	18.5	19.0	19.2	19.2	19.3	19.3	19.4	19.4	19.4	19.4	19.5
3	10.1	9.55	9.28	9.12	9.01	8.94	8.89	8.85	8.79	8.74	8.64
4	7.71	6.94	6.59	6.39	6.26	6.16	6.09	6.04	5.96	5.91	5.77
5	6.61	5.79	5.41	5.19	5.05	4.95	4.88	4.82	4.74	4.68	4.53
6	5.99	5.14	4.76	4.53	4.39	4.28	4.21	4.15	4.06	4.00	3.84
7	5.59	4.74	4.35	4.12	3.97	3.87	3.79	3.73	3.64	3.57	3.41
8	5.32	4.46	4.07	3.84	3.69	3.58	3.50	3.44	3.35	3.28	3.12
9	5.12	4.26	3.86	3.63	3.48	3.37	3.29	3.23	3.14	3.07	2.90
10	4.96	4.10	3.71	3.48	3.33	3.22	3.14	3.07	2.98	2.91	2.74
12	4.75	3.89	3.49	3.26	3.11	3.00	2.91	2.85	2.75	2.69	2.51
15	4.54	3.68	3.29	3.06	2.90	2.79	2.71	2.64	2.54	2.48	2.29
20	4.35	3.49	3.10	2.87	2.71	2.60	2.51	2.45	2.35	2.28	2.08
24	4.26	3.40	3.01	2.78	2.62	2.51	2.42	2.36	2.25	2.18	1.98
30	4.17	3.32	2.92	2.69	2.53	2.42	2.33	2.27	2.16	2.09	1.89
40	4.08	3.23	2.84	2.61	2.45	2.34	2.25	2.18	2.08	2.00	1.79
60	4.00	3.15	2.76	2.53	2.37	2.25	2.17	2.10	1.99	1.92	1.70

F values for the 1-percentage point ($P = 0.01$)

n_1 = degrees of freedom for the larger of two mean squares;
n_2 = degrees of freedom for the smaller mean square.

	n_1										
n_2	1	2	3	4	5	6	7	8	10	12	24
2	98.5	99.0	99.2	99.2	99.3	99.3	99.4	99.4	99.4	99.4	99.5
3	34.1	30.8	29.5	28.7	28.2	27.9	27.7	27.5	27.2	27.1	26.6
4	21.2	18.0	16.7	16.0	15.5	15.2	15.0	14.8	14.5	14.4	13.9
5	16.3	13.3	12.1	11.4	11.0	10.7	10.5	10.3	10.1	9.89	9.47
6	13.7	10.98	9.78	9.15	8.75	8.47	8.26	8.10	7.87	7.72	7.31
7	12.3	9.55	8.45	7.85	7.46	7.19	6.99	6.84	6.62	6.47	6.07
8	11.3	8.65	7.59	7.01	6.63	6.37	6.18	6.03	5.81	5.67	5.28
9	10.6	8.02	6.99	6.42	6.06	5.80	5.61	5.47	5.26	5.11	4.73
10	10.0	7.56	6.55	5.99	5.64	5.39	5.20	5.06	4.85	4.71	4.33
12	9.33	6.93	5.95	5.41	5.06	4.82	4.64	4.50	4.30	4.16	3.78
15	8.68	6.36	5.42	4.89	4.56	4.32	4.14	4.00	3.80	3.67	3.29
20	8.10	5.85	4.94	4.43	4.10	3.87	3.70	3.56	3.37	3.23	2.86
24	7.82	5.61	4.72	4.22	3.90	3.67	3.50	3.36	3.17	3.03	2.66
30	7.56	5.39	4.51	4.02	3.70	3.47	3.30	3.17	2.98	2.84	2.47
40	7.31	5.18	4.31	3.83	3.51	3.29	3.12	2.99	2.80	2.66	2.29
60	7.08	4.98	4.13	3.65	3.34	3.12	2.95	2.82	2.63	2.50	2.12

F values for the 0.1-percentage point ($P = 0.001$)

n_1 = degrees of freedom for the larger of two mean squares;
n_2 = degrees of freedom for the smaller mean square.

	n_1										
n_2	1	2	3	4	5	6	7	8	10	12	24
2	999	999	999	999	999	999	999	999	999	999	1000
3	167	149	141	137	135	1331	132	131	129	128	126
4	74.1	61.3	56.2	53.4	51.7	50.5	49.7	49.0	48.1	47.4	45.8
5	47.2	37.1	33.2	31.1	29.8	28.8	28.2	27.7	26.9	26.4	25.1
6	35.5	27.0	23.7	21.9	20.8	20.0	19.5	19.0	18.4	18.0	16.9
7	29.3	21.7	18.8	17.2	16.2	15.5	15.0	14.6	14.1	13.7	12.7
8	25.4	18.5	15.8	14.4	13.5	12.9	12.4	12.1	11.5	11.2	10.3
9	22.9	16.4	13.9	12.6	11.7	11.1	10.7	10.4	9.87	9.57	8.72
10	21.0	14.9	12.6	11.3	10.5	9.93	9.52	9.20	8.74	8.44	7.64
12	18.6	13.0	10.8	9.63	8.89	8.38	8.00	7.71	7.29	7.00	6.25
15	16.6	11.3	9.34	8.25	7.57	7.09	6.74	6.47	6.08	5.81	5.10
20	14.8	9.95	8.10	7.10	6.46	6.02	5.09	5.44	5.08	4.82	4.15
24	14.0	9.34	7.55	6.59	5.98	5.55	5.23	4.99	4.64	4.39	3.74
30	13.3	8.77	7.05	6.12	5.53	5.12	4.82	4.58	4.24	4.00	3.36
40	12.6	8.25	6.59	5.70	5.13	4.73	4.44	4.21	3.87	3.64	3.01
60	12.0	7.77	6.17	5.31	4.76	4.37	4.09	3.86	3.54	3.32	2.69

Appendix 26

Student's *t*

This is a 'two-tailed' table. Thus, with 10 degrees of freedom, the probability of a randomly selected value lying outside the range from −2.23*s* to +2.23*s* is 5% (2.5% in each tail of the distribution).

Degrees of freedom	*P*							
	50	20	10	5	2	1	0.2	0.1
1	1.00	3.08	6.31	12.7	31.8	63.7	318	637
2	0.82	1.89	2.92	4.30	6.96	9.92	22.3	31.6
3	0.76	1.64	2.35	3.18	4.54	5.84	10.2	12.9
4	0.74	1.53	2.13	2.78	3.75	4.60	7.17	8.61
5	0.73	1.48	2.02	2.57	3.36	4.03	5.89	6.87
6	0.72	1.44	1.94	2.45	3.14	3.71	5.21	5.96
7	0.71	1.42	1.89	2.36	3.00	3.50	4.79	5.41
8	0.71	1.40	1.86	2.31	2.90	3.36	4.50	5.04
9	0.70	1.38	1.83	2.26	2.82	3.25	4.30	4.78
10	0.70	1.37	1.81	2.23	2.76	3.17	4.14	4.59
12	0.70	1.36	1.78	2.18	2.68	3.05	3.93	4.32
15	0.69	1.34	1.75	2.13	2.60	2.95	3.73	4.07
20	0.69	1.32	1.72	2.09	2.53	2.85	3.55	3.85
24	0.68	1.32	1.71	2.06	2.49	2.80	3.47	3.75
30	0.68	1.31	1.70	2.04	2.46	2.75	3.39	3.65
40	0.68	1.30	1.68	2.02	2.42	2.70	3.31	3.55
60	0.68	1.30	1.67	2.00	2.39	2.66	3.32	3.46
∞	0.67	1.28	1.64	1.96	2.33	2.58	3.09	3.29

Index

Additivity 70, 74–75
Alkali 30, 40, 155–157
    see also Straw treatment
Allometry 104–105
Analogy, use of 33
Analysis of variance xiii, xv, 5
    assumptions underlying ANOVA 70–75
    for a change-over design 43
    for a completely randomized design 5
    for a Latin square 45
    for a randomized complete block design 5, 9, 139–140
    for a split-plot design 54, 171–173
Angles *see* Transformation
Animal-periods 41, 42
Anthelmintic 26
Antibiotic trials 24, 124, 126
Appetite 82, 129
Arc sin *see* Transformation
Areas, analysis of 173
Asymptote *see* Dose-response trials

Balance of probability 26
Bartlett's test 72, 180–182, 183–184
Behaviour 3, 120
Bent-stick models 83–86
Bias 2
Bimodal distribution 120
Binomial distribution 121–122
Blocks 3, 4–16, 139–140
    allocating animals to blocks 7
    attributes used for blocking 5
    block × treatment interaction 15
    blocking by calving date 13
    blocking by initial weight 7–9
    blocking compared with covariance 101–102
    cost of blocking 5, 6
    double blocking 10–13
    incomplete blocks 4
    purpose of blocking 4
    randomized complete blocks 4
Bootleather 25
Bovine somatotrophin 50, 65
Breeds as blocks 10, 109

Calcium 92
Calculator, pocket xiii, 139, 144
Carry-over effects 43, 44, 47, 49
Cattle
    cows 1–2, 13, 36, 44, 45, 47, 48, 49, 50, 64–65, 78, 93–94, 96–97, 102, 109, 111–112, 114–115, 116, 121–122, 167, 194–195
    heifers 13, 117–118, 197
    mastitis in dairy cows 126
    milk fat 93, 185
    milk yield 45
    steers 57
Chance *see* Probability
Change-over designs 42–51, 161–164
    when not to use change-overs 50–51
Chicks *see* Poultry
Chi-squared test 120–121, 181, 182, 201
Claims by authors 35
Coefficient of variation 32, 38, 158
    engineer's analogy 39
    estimating CV 32
    for groups 59–60
    proportional to mean 72
    reducing CV 39
Combination of experiments 127–133
Comparisons amongst treatment means 19–29, 148
    general rules 22
Completely randomized design 5
Condition scoring 116
Confounding 13–14
Constants, fitting 130–131
    parallel curves 131
    parallel lines 130

Contrasts *see* Treatment means
Copper sulphate 124–126
Correlation 6, 95, 96, 97, 99, 100
  adjacent measures 113
  correlation coefficient 97, 100
Cost–benefit analysis 26
Counts, analysis of 73
Covariance analysis 16, 95–105, 188–190, 191–192
  adjustment of organ weights 103–105
  adjustment of treatment means 99–100, 190
  as an aid to interpretation 95, 102–105, 191–193
  compared with blocking 101–102
  multiple covariance 100–101
  using preliminary variables 96–100, 188–190
Cows *see* Cattle
Cumulative treatment effects 50
Curvilinear models 27–28, 79, 86–90
  cubic model 28, 131
  exponential model 79–80, 87–88, 105, 116
  inverse polynomial model 87–88
  quadratic (= parabolic) model 27, 28, 29, 79, 86, 87, 91, 130, 186
    finding the maximum on a parabola 29
  Reading model 88–90
Cystine 69, 84, 87, 88, 89, 175–179

Degrees of freedom 5
  d.f. for a completely randomized design 5
  d.f. for a Latin square 45
  d.f. for an RCB design 5
  d.f. for treatments 5, 22
  error d.f. 5
  offsetting d.f. 11
Differences
  amending estimated difference 40
  difference between differences 138
  engineer's analogy 39
  expected between treatments 34–35, 38
Digestibility trial 41, 51, 62
  in vitro 5, 30, 62
Discrete data 119–122
Dose–response trials 20, 27–29, 78–93
  asymptotic responses 81–83
  plateaued responses 82, 87
  see also Curvilinear models
Duncan, D.B. *see* Multiple range test

Efficiency of feed utilization 102–103
Eggs *see* Poultry
Einstein, Albert 134
Energy, response to 51, 67, 82
Equality, proving it 25–26
Error 5
  amongst pens 54, 60
  in change-over designs 43
  in completely randomized designs 5
  in Latin squares 45, 46, 47
  in randomized complete blocks 5
  in split plot designs 12, 68
  within pens 54, 57
Ethics of animal research xi
Exponential responses *see* Curvilinear models

Factorial designs 62–68, 175–179
  main effects 68, 69, 175
  two for the price of one 64–66
Feed conversion ratio *see* Efficiency of feed utilization
Feed intake 129, 130, 133, 191
  feed intake response trial 193
Fisher, R.A. 22
F-ratio 6, 21, 202–203

Generality of results 126
Goats 7, 8, 9, 14, 15–16, 101, 121
  litter size 119
Gompertz function 79–80
Gonadotrophin releasing hormone 71
Grazing trials 3, 57–59
Groups of animals *see* Pens
Growth rate 5, 6, 69, 118, 191–193, 197–198

Heat stress 67, 132
Hens *see* Poultry
Homogeneity *see* Variance

Individual feeding 53, 55, 56
  records for individuals 55–56
Interaction 63–64, 75, 145–146, 175–176
  cow × period interaction 45
  two-way table 145
Inverse polynomial models 79–80

Latin squares 42–46
  analysis of 161, 165
  balanced Latin squares 45, 46–48, 167–170
    with an extra period 48–49
  combining squares 46

Least significant difference (LSD) xv, 19, 31, 137–138, 159
Least squares analysis 109–111
Library, as a refuge 40
  as a resource 33, 34
Linear model 85
  see also Regression
Litter size *see* Pigs, Sheep
Liver weight 103–104
Llamas 33
Local control 1, 3
Logarithms *see* Transformation
LSD *see* Least significant difference
Luteinizing hormone 71
Lysine 69, 175–179

Main effects *see* Factorial designs
Maintenance requirements 85, 90, 103
Mastitis 51, 126
Matrix for analysis 114, 132, 199–200
Mean square *see* Variance
Meta-analysis 127–133
Methionine 69, 84, 87, 88, 89, 175–179
Missing plots 108
Moon, landing on the 134
Mortality data 120
Multi-location experiments 123–127
Multiple range test 20, 27
  appropriate use 23

Newton, Sir Isaac 134
Non-significant differences 24, 25, 26
Normal distribution 70
Number of replicates required 31–40
  calculating numbers required 32, 33, 36, 38
  minimum numbers 39
  prudent numbers 39
  tabulation of numbers required 38
  what to do if not enough resources 39

Ockham's razor 187
Optimization 87, 88, 89, 91, 164
Orthogonality 107, 150
  orthogonal contrasts 150, 151
  orthogonal polynomials 22, 148, 149–153
    for regression analysis 152

Paddocks *see* Grazing trials
Parabola *see* Curvilinear models
Parity 5, 13, 121
Pen effects 59
Pens of animals as replicates 53–55, 124
Percentages, analysis of 74
Periods *see* Change-over designs
Phosphorus 92
Photoperiod 51, 129, 132
Pigs 27, 28, 40, 53, 54, 55, 98, 105–106, 109–110, 124–125, 191–192
  litter size in 110, 119, 158
Placebo 2
Plateau *see* Dose-response trials
Plots 4, 5
  animals as plots 5
  missing plots 108
Polynomials *see* Orthogonal
Pope, Alexander 1
Position in animal house 6
Poultry
  chicks 29–30, 56, 60–61, 69, 73, 171–174, 176
  ducks 33
  egg numbers 5, 119
  egg output 84, 87, 88, 89
  egg weight 51, 52, 86, 161–163
  hens 11, 52, 55, 56, 67, 75, 128, 129
Precision 4, 6
  in change-over trials 43
  improvement by covariance 95
  improvement due to blocking 9
  versus generality 126
Preliminary data 95, 96–102
Principles of good experiments 1
Probability 19
  balance of probability 26
  probability of success 37–38
Probit analysis 81
Proportional responses 74–75
Protein, response to 27–28, 51, 67, 78, 79
Puberty, age at 80

Quadratic equations *see* Curvilinear models

Radiation, ionizing 85
Randomization 1, 2, 3, 7, 135–136
Randomized complete blocks *see* Blocks
Rats 103–104
Reading model 88–90
Reasonable doubt 24
Records for individuals 55

Regression analysis 20, 27–29, 78–79, 95, 97, 98, 99, 162–163, 173–174, 186–187
choosing treatments for 90–92
coefficient of regression 115, 153, 156, 197–198
curvilinear regression *see* Curvilinear models
linear regression 27, 28, 83–86, 97, 98, 130, 152, 156–157, 162–163, 173–174, 186, 191, 197
for time series 115
Repeated measures 113–117
Replicates required *see* Number
Replication as a principle 1–2, 3
unequal replication 91, 107
Residual effects *see* Carry-over effects
Response surfaces 92

Sanders, H.G. 8
SAS 141–143, 147, 165–166
SED xv, 107, 108, 137, 190, 194–195
Selection of material 117, 128
SEM xv, 22, 41, 60, 137, 159, 171, 195
Set stocking 58, 59
Sexes as blocks 10, 11, 144–145
Sheep 16, 17, 26, 41, 46, 76–77, 158
litter size in 5, 119
Significance
meaning of significance 23–25
similarity to criminal trial 24
Solomon, King 26
Space allowance, response to 82, 84
Split-plot designs 12, 54, 66–68
Square roots *see* Transformation
Standard deviation xiii
Standard error 22
see also SEM
Steers *see* Cattle
Stratification 3
Straw treatment 30, 40, 62–63, 64, 155–157
Student's *t* 6, 20, 21, 204
Sum of products xv, 95
Sum of squares xv, 21
Switchback designs 49–50
Symbols xv

Technology, adoption of 36
Temperature 75, 92, 128, 131, 132
see also Heat stress
Time as a factor 42, 115–116
weekly measurements 114
Toxicity 82
Transformation of data 71
angular (arc sin) transformation 74
logarithmic transformation 71–72, 75
square root transformation 73
Treatment means, comparison of *see* Comparison
t-test *see* Student's *t*
Twinning 5, 121–122
see also Goats, litter size *and* Sheep, litter size
Two-tailed *t* tests 38

Unbalanced designs 107–110, 196
Unequal replication 107

Vaccination 43, 107
Variance 7
analysis of variance *see* Analysis of variance
homogeneity 70–71
sampling variance 114
testing for homogeneity 72, 180–182, 183–184
Vince-Prue, Dr Daphne 1
Vitamins 30, 92, 154, 155
Volatile fatty acids 16, 17, 19, 21–22, 148, 149

Weighing animals 116–117, 174
automatic weighing systems 117
gut-fill in ruminants 116

X-rays 85